T0331206

Principles of Spinning: Combing in Spinning

Principally focussed on the combing process, the initial chapters of this book explain the basic functioning of the conventional comber, which is not very different from modern combing. Various associated motions with reference to certain timings along with importance are illustrated in detail. Characteristics that set the modern comber apart are also discussed including speed and control of features such as vibrations and fatigue. The role of modern electronics in mechanism control, including related calculations and solved examples, is described.

Features:

- Exclusive title focussing on various elements of combing in spinning.
- Includes detailed functioning of conventional and modern combers.
- Explains various motions in combing pertaining to different timings.
- Discusses the role of modern electronics in controlling the mechanisms and offering online controls.
- Features solved examples at the end to tackle problems at the shop-floor level.

This book is aimed at senior undergraduate/graduate students, textile process engineers and related manufacturing technologists, quality assurance professionals in textile engineering, staple fibre processing, spinning of staple fibres, textile marketing/design, textile technology, yarn manufacturing, spinning machines and design of yarns.

Principles of Spinning: Combing in Spinning

Ashok R. Khare

CRC Press
Taylor & Francis Group
Boca Raton London New York

CRC Press is an imprint of the
Taylor & Francis Group, an **informa** business

Designed cover image: Ashok R. Khare

First edition published 2024
by CRC Press
6000 Broken Sound Parkway NW, Suite 300, Boca Raton, FL 33487-2742

and by CRC Press
4 Park Square, Milton Park, Abingdon, Oxon, OX14 4RN

CRC Press is an imprint of Taylor & Francis Group, LLC

© 2024 Taylor & Francis Group, LLC

ISBN: 9781138596597 (hbk)
ISBN: 9781032534909 (pbk)
ISBN: 9780429486555 (ebk)

DOI: 10.1201/9780429486555

Typeset in Times
by Deanta Global Publishing Services, Chennai, India

Contents

Contents

Preface

My earlier venture in this field had been a modest attempt to write some useful information for textile students. It was because I then found that such information was scanty and students have to dig for the required information regarding the subjects from different sources. Even then, I must admit that I received tremendous responses from students all over India and everyone categorically mentioned that the efforts were very useful for the students.

It is after almost 12 years that I thought of rehashing the whole format and at the same time, adding some very useful information, which I am sure will again be welcome by the student community.

While maintaining the basics of combing, all the figures are redrawn to be close enough technically. In almost every chapter, there have been sincere efforts to refer to the techniques used in modern combers. As usual, the solved examples at the end would help the students to get a fair idea of how to tackle the problems at the shop-floor level

Like this book, I am again thinking of rehashing the information on various other topics in spinning such as – blow room, carding drawing, fly frame and ring frame. All these would come in the course of time in an almost similar manner. Lastly, I am sure that the student community will find very handy and useful information in this book.

Dr. Ashok R. Khare

Acknowledgements

Since I was a student in college, I had always dreamt of becoming a teacher. A few of my teachers were my idol then. I feel proud to mention their names. This is because, later when I became a teacher in the same college after several years, I always got inspired by remembering them, their style and philosophy of their teaching, their sincerity and devotion to the profession and their skill in making the difficult things look simple. All of them had good industrial experience and were, therefore, able to share their knowledge with the students.

Prof. D.B. Ajgaokar who later became the first Principal of D.K.T.E. Institute Ichalkaranji, encouraged me to venture into writing the books for students. After my first venture, he continued his encouragement and support for writing more. This is how I was able to take up this vast task of writing a book series on spinning technology.

The late Prof. M.K. Naboodiri had been my philosopher and guide and very few would really know that I had family-like relations with him. When any problem was posed to him, he would never offer haphazard answers. He would meditate and then come out with a logical solution.

The late Dr. V.S. Jayram was my first teacher when I first started learning about textile technology, a field which was unheard till the completion of my school days. I used to immensely like the way he taught spinning. I must admit that the credit for my own continuing with textile studies solely goes to him. As a student, I never used to miss his lectures. I must also confess that his style of teaching inspired me to come back to my college as a teacher. When I requested him to edit this volume, he very gladly accepted and completed the work. But most unfortunately, within a short period thereafter, he passed away. I will always remain indebted to Dr. Jayram

Dr. S.G. Vinzanekar, my mentor during my whole career as a teacher, was always a source of inspiration for giving me several opportunities to learn things. When he would entrust any job, he would always fully back and support me, irrespective of whether it was a success or otherwise and even in odd circumstances. He was my guide for my Ph.D. work and there too, he became my senior philosopher friend.

I am extremely thankful to the Director of CIRCOT (Central Institute for Research on Cotton Technology), Mumbai, India, for helping me from time to time, to make the CIRCOT research work on cotton technology available. The thanks are very much due to two giant and reputed machinery manufacturers – Trumac–Trutzschler and Rieters for providing me with beautiful diagrams supporting the theory of blow room machines. Without their help and permission, this book would not have been what it is now.

The base of this book appears to be similar to the Manual of Cotton Spinning book series published long back by Shirley Institute, though in treatment to the subject information, it differs. Even then, I profusely thank The Textile (formerly Shirely) Institute, Manchester, London, UK, for giving me the inspiration to write and add some useful information to the ocean of textiles. Equally important was the permission from Elsevier (former Butterworth publication) who permitted me to refer to their book *Spun Yarn Technologies* (by Eric Oxtoby). I am greatly indebted to them. I hope, in its present form, it is still useful to the students. The thanks are also due to my well-wishers, who directly or indirectly helped me in this venture.

Last but not least, I would be failing in my duties if I do not mention the name of my father, the late Shri Ramchandra Narayan Khare, who since my childhood, groomed me to become a good student, a good teacher and a good citizen. He himself was a born teacher and expert in child psychology. When I was thinking of leaving my job in industry to join a teaching career, he asked me only one question – 'would you take-up this career with all sincerity and dedication?', He also imbibed in my mind that if I have to become a good teacher, I always need to be a good student. I have never forgotten his words during my whole professional career as a teacher.

If I have authored this book, the credit goes to all these great personalities; in some way or another, they have been instrumental in making me as I stand today.

Dr. Ashok R. Khare

About the Author

Ashok R. Khare is a graduate, post-graduate and doctorate from a well-known technological institute – Veermata Jijabai Technological Institute, Mumbai, India (formerly known as Victoria Jubilee Technical Institute). He graduated from this Institute in 1970 and went on to serve a well-known textile group of Mafatlal mills. After serving for nearly five years in the textile mills, he returned back to his Alma Mater in 1975 as a lecturer in textile technology. In due course, he was promoted to assistant professor and professor.

In the last phase of his service in V.J.T.I. he took over as the Head of the Textile Manufacture's Department. Almost during the same tenure, he held the position of deputy director in the same institute. He has written several articles on card cleaning efficiency, the role of UniComb and extended research work on the influence of doubling parameters on properties of blended doubles yarns.

1 Combing Preparation

1.1 COMBING PREPARATION[1]

1.1.1 SHORT FIBRES

With cotton, it is sometimes more interesting to find the fibre length distribution rather than the absolute values of important individual fibre parameters such as effective length, mean length and short fibre content.

Sophisticated instruments such as AFIS or HVI reveal the values of these parameters but the fibre distribution diagram given by the Baer sorter is unique in this case. Though time-consuming, it gives the representation of fibres of varying lengths and their proportion in the cotton.

As shown in Figure 1.1, it can be seen that there is a certain proportion of fibres (OAK'K) longer than the Effective Length (line LL'). Though this proportion is small, the length alone outweighs the number. Around effective length, there are a sizable proportion of fibres representing the majority of the fibres (say C). Equally interesting are the fibres which are shorter than even half the effective length (shaded portion – RBR' – (**B**)). These fibres are termed '**short fibres**' (S.F.). The proportion of these short fibres decides the value of another important parameter – '**mean length**' (M.L.) – line MM' in Figure 1.1.

It will be of some interest to know how these different parameters are found after the Baer sorter diagram is constructed. Point Q is the midpoint of line OA. The line QP' is drawn parallel to the 'x-axis'. Thus P' is the point where it meets the Baer sorter diagram. From P', a perpendicular line is drawn to meet the x-axis at P. The line OP is divided into four equal parts (not shown). The upper quarter is OK. A perpendicular line is drawn from K to the x-axis in a vertical direction to meet the curve at K'. Point S is the midpoint of the line KK'. A line parallel to the x-axis through point S meets the graph at R'. A perpendicular from R' meets the x-axis at R. The shaded area R'RB represents the proportion of **short fibre** in the tested sample.

The line OR is again divided into four equal parts. The upper quarter is line OL. A perpendicular in the vertical direction from point L meets the graph at L'. Thus, LL' represents the '**upper quartile length**' or '**effective length**' (E.L.). Similarly, line MR is the '**lower quartile**' of line OR. The line MM' is referred to as '**mean length**'.

It is well known that the fibres of categories (1) longer than the effective length (e.g., LL') and (2) 'C' being the majority of fibres around effective length are very useful for spinning them into yarn. However, short fibres (B) invariably pose major problems in processing and lead to higher yarn irregularity. By a definition, a fibre is 'short' if it is less than half the effective length of the lot. However, it may be noted

DOI: 10.1201/9780429486555-1

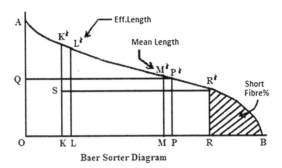

Baer Sorter Diagram

FIGURE 1.1 The Baer sorter diagram:[2] It is very useful while carrying out combing operations. It helps in deciding the level of comber noil to be extracted.

that the term 'short fibres' is only relative and is with reference to the effective length of the lot being processed. Thus, for cotton with E.L. 30 mm, the fibres shorter than 15 mm may form a category of short fibres; whereas with E.L. as 24 mm, fibres shorter than 12 mm may only be considered as short fibres. The term 'short fibres' is not the inherent property of fibres, but it is associated with and related to the E.L. of the mixing. Thus, if a certain proportion of fibres is termed as short fibres for finer mixing, a part of it, can very well form an important portion of comparatively lower-grade mixing.

1.1.2 PROBLEMS WITH SHORT FIBRES

After the sliver formation in carding, the subsequent processes up to the spinning of yarn (except combing) invariably include the process of drafting for parallelization and uniformity, and then systematic attenuation (thinning-out) of fibre strands up to the yarn stage. The movement of the fibres during drafting has to be more precise and uniform; otherwise, a typical defect called a 'drafting wave' occurs. The presence of short fibres prohibits this uniformity as their movement cannot be precisely controlled during drafting. The drafting wave is the formation of thick and thin places in the drafted strand and thus leads to higher non-uniformity or irregularity. Apart from not contributing much to the strength of the yarn, these short fibres are also not bound firmly into the body of the yarn. They, therefore, protrude out and lead to 'hairy' yarn, a serious defect in fine-quality yarns and fabrics. Further, a significant proportion of these short fibres is not fully matured and usually has a high potential to form neps.

1.1.3 OBJECT OF COMBING[2]

The combing operation is mainly carried out to remove such fibres which are shorter than a certain pre-determined length. As seen earlier, removing these shorter fibres also removes the trouble they are likely to cause during drafting. The strand of fibres becomes quite free from their presence and the subsequent operations of drafting and thinning up to the spinning stage are improved. This results in a more regular and uniform product. Ultimately, the yarn produced is not only far more uniform and much less hairy but also has a better appearance, especially in terms of a substantial reduction in neps.

It is often said that the process of combing 'upgrades' the cotton. It means that the product that is formed after combing is superior in quality. It also means that the same combed material, within reasonable limits, can be spun profitably to finer counts. In this case, it may be possible to spin the finer yarn without greatly affecting the quality. In mills, either one of the objectives is sought or else, both are combinedly achieved.

Let us assume that the two yarns are spun from the same mixing. But one of them is from combed material and the other one is from carded material (not combed). When the properties such as strength, uniformity, appearance, etc., of the above two yarns are compared, it will be realised that the combed yarn is superior in these aspects. Equally possible will be that when one yarn is spun from the combed sliver material to a slightly finer count and the other, from the same mixing, but not combed, is spun to a somewhat coarser count, the two yarns may appear to be similar in some of these properties. This is what is meant by upgrading the cotton.

The typical nature of the combing process involves a fine set of a number of rows of needles (Figure 1.2) passing through a fringe of lap held firmly at its other end. The needles in the rows thus pick-up short fibres which loosely linger at the front portion of the fringe presented to them. Being short, they are also not firmly held along with the other longer fibres. It is also possible that, along with short fibres, the rows of needles pick-up a significant proportion of neps, fine kitties, leaf particles, etc., from the fringe. Some cleaning is therefore possible.

It may be mentioned that the cleaning process of comber is quite different in its character from that carried in either blow room or card. The beater blades, spikes,

Needle Rows

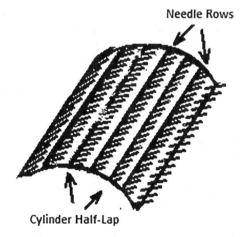

Cylinder Half-Lap

FIGURE 1.2   The cylinder half-lap:[1,2] It consists of needle rows neatly arranged and mounted on a segment. The needles become progressively finer to thoroughly comb the material presented to them.

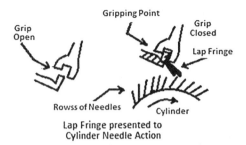

Lap Fringe presented to
Cylinder Needle Action

FIGURE 1.3 The gripping of the lap:[1,2] The typical shape of the top and bottom nippers helps in the firm gripping of the lap fringe.

pins or metallic saw tooth wires in the blow room or card are used to open out the clusters of cotton and simultaneously clean the stock, with the help of the grid bars. In combing, however, the. needles, with their fine points, do this operation by themselves When one end of the lap fringe is held firmly (Figure 1.3) and its other end is presented to the rows of needles, there is a disentanglement of fibres in the strand owing to the action of a set of parallel rows of needles. The action of the needles straightens out the fibres in the fringe and aligns them parallel to the direction of the movement of needles, with their fine points, do this operation by themselves. When one end of the lap fringe is held firmly (Figure 1.3) and its other end is presented to the rows of needles, there is disentanglement of fibres in the strand owing to the action of set of parallel rows of needles. The action of the needles straightens out the fibres in the fringe and aligns them parallel to the direction of the movement of the needles.

The combed-out material has thus better fibre orientation, alignment and parallelization, all enhancing the lustre. There is yet another advantage of the action of needles. While passing through the front end of the projected lap fringe, the needles straighten out the bent fibres (fibres with leading hooks). It may be also mentioned that when the fibres are finally being detached, it is again possible that the bent fibres (fibres with trailing hooks) are also straightened out by another contrivance called 'top comb'.

This action helps in improving the fibre extent and thus refines the length parameter of the fibres thus contributing to building up yarn properties, especially the yarn strength. As can be seen from Figure 1.4, the fibre extent AB (L_1 cm) and the hook extent BC (L_2 cm) are the two components of the length of the fibre in its hooked condition.

Thus, during processing, only the fibre length L_1 is utilized. The length L_1 is utilized when the fibres lie in their hooked condition in the strands constituting a yarn. After the combing operation, the fibres are straightened out and the length of the fibre becomes AC ($L_1 + L_2$). With the above discussion, the objectives of combing can now be summarized as follows:

1. To remove the short fibres below a pre-determined length and upgrade the cotton.
2. To straighten out the fibres and improve the fibre extent.

Straightening of Leading Hooks

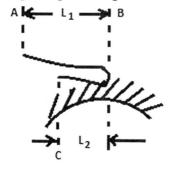

During Combing

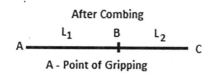

After Combing

A - Point of Gripping

FIGURE 1.4 The straightening of hooks:[1,2] The needle, while passing through the hooked limb, straightens it, thus increasing fibre extent.

3. To remove neps and other foreign matter such as kitties, leaf or trash particles, etc.
4. To improve the general appearance of the material.

REFERENCES

1. *Manual of Cotton Spinning: "Draw Frames, Comber & Speed Frames": Frank Charnley*, The Textile Institute Manchester, Butterworths, 1964
2. *Elements of Cotton Spinning: Combing: Dr. A.R. Khare*, Sai Publication

2 Combing Preparation

2.1 COMBING PREPARATION

2.1.1 Earlier History[1]

The first Heilman comber introduced in 1846 had only one head* and was fed directly from a sliver creel. The sliver preparation was modified later and four or more slivers were used to form a lap sheet, suitably wound onto a wooden bobbin and then fed to the combing machine.

A Derby doubler was another machine popularly used in 'double combing', a process involved in combing cotton with a high level of noil extraction. It enabled the assembling of comber slivers into a lap which was again fed to a comber for a repeated combing operation to extract more noil.

This machine merely gathered the fed slivers and made again the lap, suitable in width for the comber to reprocess. The Derby doubler, however, had no drafting element. The combination of a present-day 'Draw Frame Lap Former'[†] is almost similar to the pre-comb Draw Frame and Derby doubler, the sequence used then. Subsequently, a drafting element was added to the Derby doubler to give a drafting range of 1.7–2.0 and this is how the present 'Sliver Lap' (@) machine has come into vogue.

Dobson and Barlow patented an improved version of the Draw Frame in 1874. In this, six, or any convenient number, of drawing heads were placed side-by-side and each head was creeled with slivers from card cans. The drawn slivers from each head were guided over a curved plate to turn their direction at 90°, onto a horizontal plate, and were later sandwiched and wound into a lap. This became, in later years, another machine called a 'Ribbon Lap' machine (#).

The Ribbon Lap machine was very well received and adopted, especially after the introduction of the Nasmith comber. The weight of the lap demanded then, was not very heavy then and the sequence of Sliver Lap and Ribbon Lap machines enjoyed popularity till almost 1950.

The next positive step in combing was taken around 1952 when the concept of high production combing was first introduced by Paul Whitin. In this, a totally new lap preparation was introduced along with high production rates at the comber. It involved making heavier laps which were expected to work satisfactorily on the modern Whitin comber.

Around the same time, the pioneering work carried out by Morton[5] revolutionized the idea of the configuration of fibres in the card sliver. This forced the idea of making a comber lap to change with three distinct objectives – the Degree of Parallelization, lap regularity and the direction of presentation. Parallel research also revealed that there

---

* Head is that section of the machine through which a fully combed sliver is delivered.
[†] See Page 25 @ See Page 17 # See Page 24

DOI: 10.1201/9780429486555-2

was a certain minimum pre-comb draft and number of doublings which were necessary to give a comber lap a certain level of fibre orientation, disentangling of fibres, correct direction of feed and lap uniformity. All incorporated, the new Whitin lap preparation along with high-speed combers became the most popular comber system.

2.1.2 FIBRE ARRANGEMENT IN THE CARD SLIVER[1,3]

Morton and Summars[5] were the first to examine the fibre arrangement in the card sliver. Since then, it has been known that the fibres in the card sliver are not parallel to the line of delivery of the sliver. In fact, they are highly crisscrossed. It is owing to this arrangement of fibres that the inter-fibre cohesion, as a result of fibre entanglement, helps in the easy doffing of the web from doffer in the card. As an outcome of this research work, two important parameters were recognized – (a) Fibre Extent and (b) Fibre Orientation. The fibre extent is defined as the projected length of the fibre along the sliver axis. The ratio of fibre extent to fibre length is called the 'Degree of Parallelization'. In the card, it is found to be between 0.54 and 0.59.

When the fibre orientation was studied, it was found that almost half the fibres (50%) had 'trailing' hooks (and hence called the 'majority of the hooks'), whereas; only 15–20% of the fibres had 'leading' hooks (called the 'minority of the hooks'), the remaining being hooks at both ends and those belonging to miscellaneous category.

With the fibres in the card sliver lying so randomly and haphazardly, it was certain that such material, when presented to the comber, would cause excessive strain on the cylinder needles, thus resulting in damage. It would also cause damage to the fibres and a considerable amount of spinnable fibres would go into waste (or noil) as they would be simply treated as shorter fibres during the combing process.

Ultimately, the comber would have done more harm than improving quality. Therefore, fibre presentation and fibre parallelization before combing are the most important aspects one has to consider.

When these two factors are taken care of, the fibre cohesion owing to high entanglement in feed material to the comber is substantially reduced. This is owing to fibre parallelization achieved in the pre-combing process. It allows the fibres to be detached more freely without much disturbing the other neighbouring fibres in the comber lap during the combing operation. In a way, it also controls the amount of waste extracted in the combing process and thus improves the 'fractionating efficiency'* of the comber. The fibre presentation involves the use of a number of combing preparatory machines.

2.1.3 FIBRE PRESENTATION[1]

The fibres F_1 and F_2 in adjoining Figure 2.1, have the leading hooks. These hooks are very likely to be straightened out due to an action of cylinder needles (line CD moving in the direction of arrows).

During this straightening, the fibre extent is automatically improved. As against this, the fibres F_3 and F_4 have hooks at their trailing end. In this case, however, the

---

* An ability of the comber in effectively removing the shorter fibres without allowing the longer fibres to go into waste. [See chapter – 'Fractionating Efficiency']

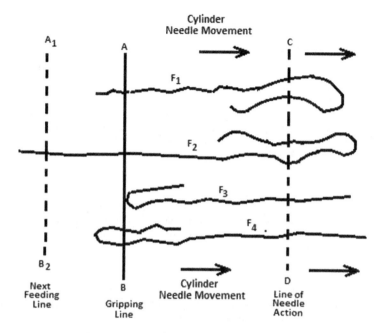

FIGURE 2.1 Removal of fibres in combing:[1,3] The cylinder needles while passing through the fibre fringe, straighten the leading hooks. However, the needles remove those fibres which are not held by the nippers at the time of combing.

needle movement is not able to straighten these hooks. It can also be seen from the same figure (Figure 2.1) that F_3 in its present position, will simply be carried away by the needles during combing.

As regards the fate of the fibre F_4, it will remain intact in the present cycle. In the next cycle, however, its hooked portion will be brought ahead to lie between B and C and therefore, its fate will be similar. Being not held by the gripping line, the needles will simply take it away to form waste. As stated earlier, a card sliver has a predominance of trailing hooks and it almost amounts to 50% of the total fibres. Thus, it would be of great advantage to present this majority of hooks as leading hooks to the action of cylinder needles. It can be also seen from the Figure 2.1 that along with the direction of the needle movement, the direction of the material feed entering the comber is also the same (the feed line moving from line A_1B_1 to line AB. Thus, it is conceivable that the majority of the hooks in the material presented to the comber would continue to remain as the "**leading direction**".

2.1.3.1 Hook Reversal[1,3]

It has been already mentioned that, as delivered by the card, the majority of the hooks in the card web are in the trailing direction. However, when the card sliver is coiled into the card can and the same is subsequently fed to the drawing frame, the direction of the majority of the hooks is reversed. This means that every time the

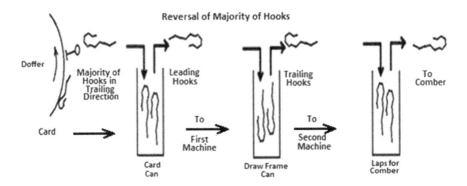

FIGURE 2.2 Reversal of hook direction in post-card process:[1,3] After coiling the sliver into the can or winding it in the lap form, when the material is either withdrawn from the can or unwound from the lap, there is a reversal the direction of hooks Thus, the leading hooks become trailing and vice versa.

sliver material is coiled into the can or wrapped around a spool, and then withdrawn at the next process, the direction of the sliver changes. This also changes the direction of the majority of hooks (Figure 2.2).

Thus, there is a reversal of hooks, i.e., trailing hooks change to leading hooks and vice-versa at every process involved after carding. It can be therefore realized that immediately after the card, the first process involved would have the majority of the hooks in the leading direction.

But it has been already stressed that the fibres in the card sliver are most randomly disposed of, it would not be advisable to feed the card sliver directly to the comber. Also, the feed material for the comber needs to be in the lap form and hence, some sequence of preparatory machines is necessary. It can, therefore, be stated that, after the card, every third or fifth machine could be a comber. In short, there needs to be an "even number of machines" in between the card and the comber in the combing preparatory sequence.

2.1.4 Fibre Parallelization[1]

For effective and smooth combing action, it is necessary to remove fibre entanglement and make them parallel to the material axis (in the case of the lap – the material plane). The process of combing by machine is analogous to combing that women do with a hair-comb. When the hairs are highly entangled, it is very difficult to move a comb through. When the comb is forcibly pushed through, in this condition, some of the hairs are likely to be uprooted. Without putting the comb more deeply, therefore, the ladies superficially put the comb into their hair, lightly push it through their hair and remove the entanglement initially. Similarly, it is very important that the initial crisscrossing at the card sliver stage has to be improved by disentangling the fibres before combing them. The parallelization achieved through pre-combing preparatory processing has this beneficial effect. It reduces the strain on the fibres during combing. It also reduces the strain on the combing needles which, in reality, are merely soldered to each of the strips representing rows of needles.*

Further, the needles are very fine, sharp and delicate and hence the strain due to fibre entanglement is equally experienced by them. As a result, it is likely that during combing if the needles have to pass through a highly entangled lap under force, the needles would either get damaged or would be totally uprooted. The draft is given in the machines which prepare the lap for the comber. This helps in reducing the entanglement and tries to lay the fibres parallel. The parallelization depends upon the amount of draft given in the combing preparatory process.

When the level of this parallelization is insufficient or inadequate, some of the areas in the comber lap have high entanglement and it disturbs the combing process. This is because; the fibres in these areas create high frictional forces.

It can now be seen (Figure 2.3) how the trailing hooks can be the source of trouble during combing. Let us assume that line MN is the gripping line for the fibre fringe after the combing operation; whereas line PQ is the gripping line for the comber lap before the fringe is subjected to combing by cylinder needles. The next feed line for the comber lap is the line ST which advances through a certain length (say 'x') during each combing cycle.

With this advancement (distance x), the fibres such as A, B and C would go out of the lap gripping line PQ during the subsequent combing cycle and would be removed. This is because, in spite of them being long enough, they would move ahead of the gripping line PQ. Thus, during cylinder combing, these fibres would be simply removed. Therefore all such fibres with trailing hooks, as they are not gripped under the nipping line PQ would go to waste. The fibres of the categories D, E and F have their trailing end extending behind the next feed line. Their fate would depend upon whether they are pulled by detaching rollers during detachment

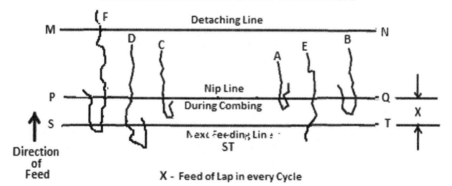

Position of Fibre Hooks & their ultimate Fate

X - Feed of Lap in every Cycle

FIGURE 2.3 Fibres under the gripping line:[1,3] The position of the fibre and its hooked end under the nipper at the time of combing decide whether its hooked end is straightened and also whether the fibre stays as fibre or is removed as waste.

* On all modern combers there is saw tooth segment (Unicomb) on the cylinder.

operation. Otherwise, they would be still under the nip line in subsequent combing cycles. The parallelization in the combing preparatory process helps in reducing the total hook content and thus improves the fibre extent. With a proper sequence of combing preparatory machines, the trailing hook content in the comber lap can also be reduced. This would improve the fibre extent. Thus, the chances of survival for all such fibres during combing are greatly improved.

2.1.5 PRE-COMB DRAFT AND NUMBER OF DOUBLINGS[1]

Fibre parallelization in any spinning process is mainly due to the draft employed (except the fringe parallelization due to cylinder needle action). In combing, fibre parallelization is achieved through the pre-combing processes. With a higher draft, there is better fibre parallelization, fibre orientation and improved fibre extent, all of which, again, significantly affect the level of waste extraction. This draft level depends upon the type of material and its staple length. It is found that short-staple cotton is more easily parallelized than longer-staple cotton. It is seen from the graph (Figure 2.4) that, as the pre-comb draft increases, the waste extracted at the comber reduces (without changing any other parameter).

What differs, however, is the nature of the graph. The graph also differs when the experiments are carried out with two different types of cotton (curves M and N). With curve M, the fall in the waste is continuous up to point B; as against this, with curve N, the decrease in the waste is initially sharp, but up to point A. The influence of the pre-comb draft in these two cases is thus different (points A' and B'). Thus, with a comparatively less pre-comb draft (point A – graph N), the reduction in the comber waste is comparatively more (N$_1$ to N$_2$). However, when the draft is increased beyond this point, the corresponding reduction in the comber waste becomes steadily less pronounced.

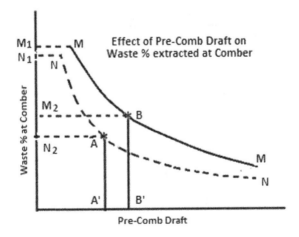

FIGURE 2.4 Pre-comb draft versus comber noil:[1,3] The pre-comb draft decides – the level of fibre parallelization, the number of hooks present and the reduction in the hook extent in the comber lap. Thus, it has a direct bearing on the noil extracted at the comber.

Similarly, with the other cotton, the reduction in the level of comber waste is more pronounced but only up to point B. These two curves, as mentioned earlier, represent the cotton of two different classes and staple lengths. The magnitude of the pre-comb draft beyond points A and B in these two respective cases, if it involves an additional machine in the combing sequence, will simply involve additional processing costs which may not be compensated by further savings in comber waste. An excessive pre-comb draft may also pose yet another problem – lap licking.* This is due to excessive parallelization resulting from a higher draft and making the comber lap softer.

The excessive parallelization in the pre-comb stages makes the sliver very weak, and there are higher creel breaks in lap-making machines. The sliver withdrawal, in this case, needs to be assisted by careful handling of sliver in creel so as to avoid any false draft which is unintentional. This draft is experienced by the material while passing over the rough surfaces.

Self-Cleaning Effect:[6] During every detachment, only 16–18% of the fibres are drawn from the thick lap sheet. The remaining fibres in the lap sheet offer some kind of resistance to hold back neps and other foreign matter. Most of these come to the tip of the subsequent fringe presented to cylinder combing action in the next cycle and easily get removed by the cylinder needles. This is called the 'self-cleaning effect'. This again depends upon the level of parallelization attained by the lap sheet during pre-combing preparation. As mentioned earlier, the more the parallelization, the greater the saving in the comber noil, but the lesser the self-cleaning effect. It may be mentioned here again that the savings in the comber noil are accrued without change to any of the comber settings and also do not affect the yarn quality.

Thus, more parallelization reduces the self-cleaning effect – the one responsible for restricting neps and other impurities from passing into combed sliver. The self-cleaning effect is inversely proportional to the pre-comb draft. When the yarn is spun with a higher pre-comb draft, there is no appreciable increase in yarn strength. However, the yarn's cleanliness is seriously affected. Further, extra parallelization in the pre-combing stage leads to lap splitting as the fibre-to-fibre adhesion is markedly reduced. It is also observed that with a higher pre-comb draft, there is a marked increase in the hairiness of the lap.

2.1.6   COMBER LAP PREPARATORY MACHINES[1,3]

The lap preparation for comber consists of a sequence of machines. As mentioned earlier, this sequence has two machines (an even number of machines). Employing more machines in multiples of two is also possible, but highly uneconomical. This is because, it entails not only additional cost of machines, labour, space and power but also does not give proportionate returns in terms of saving in comber noil.

The conventional sequence for comber lap preparation is Sliver Lap and Ribbon Lap machines. Both are, in a way, lap-making machines. A modified version of the above sequence is a Draw Frame and Sliver Lap machine. There is another parallel sequence – Draw Frame and lap former –introduced by Platt. A modern sequence meeting both the technological and the production requirements needs to incorporate

---

* The adjoining layers of the lap sheet, sticking to each other during unwinding.

the necessary pre-comb draft and doublings and produce laps of higher weight that the modern comber is capable of handling. Whitin and Sons were the first to introduce the modern pre-comb machinery sequence – the Whitin Draw Frame and Super Lap Former, along with their then high-speed combers, some 60 years back.

2.1.6.1 Sliver Lap Machine

This is a conventional machine and has the combined features of both, the drawing frame and lap-making part. Thus, the first half up to the drafting rollers almost resembles the draw frame; whereas the second half is a lap-making unit. It may be mentioned here that the drafting system, unlike the one on a conventional draw frame, plays a different role and therefore the drafting capacity of the system is very limited.

The sliver from card cans (Figure 2.5) are kept behind the machine and the slivers from these cans are guided through a single preventer roller or lifting roller system. There are usually 20–22 card slivers fed at a time to the sliver lap machine. The slivers are passed over suitable guides and led through the nips of a single preventer system (Figure 2.6). The function of the lifting or single preventing system (Figure 2.5 (A)) is basically to see that sliver breaks in the feed zone are sensed and detected. Immediately, thereafter, the machine is stopped. The functioning of this sliver break-stop motion is very simple. Both the top and the bottom rollers are given electrical connections. The bottom roller is continuous and extends over the full width of the machine; whereas the top rollers are split into pairs of two. As long as the slivers run through the nip, there is no contact between the top and bottom rollers. But when any of the slivers breaks, the respective top roller makes the contact with the bottom roller and thus completes the electrical circuit which ultimately stops the machine.

Much earlier versions of this machine had mechanical sliver break sensing and stopping arrangement where, in place of a single preventer roller, there used to be

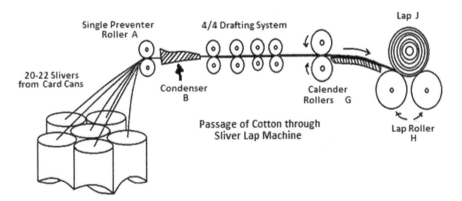

FIGURE 2.5 Material through Sliver Lap machine:[1,3] The basic idea of using this machine is to combine slivers to make a lap of suitable weight per unit length with some level of uniformity both across and along the length.

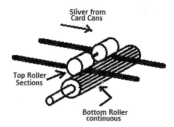

FIGURE 2.6 Single preventer roller:[1,3] When the required number of feed slivers is decided to make a lap of a certain weight per unit length, it is very important to stop the machine when any one of the feed slivers break.

sliver guiding spoons (Figure 2.7). The slivers, in this case, were guided over these spoons and then led on to lifting rollers. The lifting rollers, in this case, were merely made to act as a device to lift the slivers; whereas the sensing of the sliver break was done by the spoons. With the slivers continuing their journey over the spoons and in the process pressing them a little, the tail end of the spoon remained lifted up.

When the sliver broke, the tail end of the spoon used to turn around the pivot edge and fall down. The tail of the spoon used to be then picked by the vibrating bar on its way. This finally resulted in stopping the machine. The great disadvantage of the spoon-type of motion was that being purely mechanical, there was a marked delay in sensing the sliver break and conveying it further to the shifting of the main driving belt to stop the machine.

Further, its functional efficiency also depended upon the smooth turning over of the spoons when the sliver broke. Therefore, as compared to the mechanical sliver

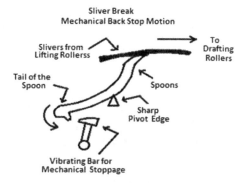

FIGURE 2.7 Sliver break sensing spoon:[3] One of the great limitations of a purely mechanical system, was that both the sensing and relaying of the signals to stop the machine were not instantaneous. This was overcome by an electrical system where sensing as well as relaying was much faster. It helped in improving both the machine efficiency and sliver uniformity.

break-stop motion, electrical sensing and conveying, as done by the single preventer roller, is much faster and almost instantaneous. It may be noted here that the electrical sensing system assumes one very important thing and that is the non-conductivity of cotton material passing through sensing preventer rollers. This is true when the moisture content in the cotton material is lower. But with higher humidity in the department, especially in rainy seasons or wet nights in winter, the moisture content in cotton also increases and it partially starts conducting the electricity. Hence, in such situations, faulty sensing of the absence of a sliver is registered and the machine frequently stops. In extreme cases, the worker operating the machine finds the machine stopped even when there is no sliver break at the creel. If this is repeated frequently, he feels irritated and thus tends to tamper with the electrical connections, which again is worse.

The condenser (B) (Figure 2.5 and 2.8) has several guide plates which provide channels for the slivers coming through a single preventer system. The basic object of the condenser is to restrict the width of the slivers as they enter the drafting system.

This is essential as not only the slivers are laid side-by-side before they enter the drafting system, but also the width to which the slivers are restricted decides the width of the lap sheet that the slivers are going to form after they emerge from the drafting system. This gives the precise width to the lap formed later. While guiding the slivers over the condenser, care must be taken to avoid any folding or overriding of slivers which form the sheet; otherwise, it will lead to uneven thickness of the lap sheet.

The drafting system placed subsequent to the condenser consists of traditional four-over-four rollers. As compared to the drafting system on the draw frame, the pressures on the top rollers in the sliver lap machine are much higher. This is because

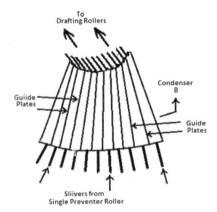

FIGURE 2.8 Condensing of slivers:[3] While making the comber laps, their width is one of the important dimensions. The sliver condenser decides this width by suitably converging the feed slivers.

the material controlled and processed during drafting is much heavier. The draft employed, however, is limited to only 1.5–2.0.

Thus, there is not much drafting and hence parallelization is not the main object of this machine. The basic purpose of this low draft is mainly to provide a little flexibility for small adjustments to obtain the desired lap weight per unit length. It also provides some loosening of slivers so that when they come out of the drafting system, there is a good merging of adjoining slivers with their neighbours.

This helps in obtaining a little more homogeneous lap structure. This again is very important as the slivers lose their separate identity that they have before drafting. It results in a well-textured lap.

The amount of draft that the machine can use is very low, only a little adjustment in the lap weight is possible. Hence, when a major variation in the lap weight is required, it is customary to vary the number of slivers fed to the machine. Here again, too much reduction in the number of slivers to arrive at the required lap weight is likely to affect the lap uniformity owing to variation in the number of doublings. It is therefore advisable to suitably produce the appropriate hank of card sliver so that by keeping the same number of doublings, the required lap weight can be arrived at. The drafted sheet in the form of a thick web comes out of the front pair of drafting rollers. It then passes through a nip of pair of calender rollers which are bigger in size and are heavily loaded. The sheet is compressed to consolidate the sliver lap and is made more compact (Figure 2.9). This is essential as the lap when wound in this sheet form has to be free from any lap-licking tendency during subsequent unwinding.

A compact lap sheet also helps in reducing the full-lap volume for a given lap weight. However, when this machine is used in the traditional sequence (Sliver Lap – Ribbon Lap), a very high pressure to consolidate the lap is also not beneficial. This is because first the lap formed on the Sliver Lap is not the final lap. Second, this lap

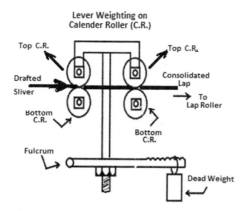

FIGURE 2.9 Mechanical loading on calendar rollers:[3] The lap made for the comber has to be compact and well compressed. This avoids its splitting or licking of the lap in combing. The calendar rollers in lap-making machines are therefore heavily weighted.

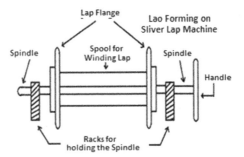

FIGURE 2.10 Mounting of spool on lap spindle:[3] The spools are held by the lap spindle which in turn is guided by the racks. The spools are supported by lap flanges on either side so as to restrict the width over which the lap is wound.

is again required to be drafted at the Ribbon Lap, the next machine in the sequence. Nevertheless, the compacting pressure should be adequate enough so as to avoid any lap licking at the feed end of the Ribbon Lap machine. The lap is wound on wooden or metallic spools (Figure 2.10). The spool is hollow and a lap spindle is made to pass through it. As the width of the comber lap, in comparison to that made in the blow room is much smaller; the lap spindle is also much shorter in length.

Whenever the lap is full, the spools are required to be replaced. At this time, by using the turning handle (A' of the handle to be engaged with A of the rack pinion – Figure 2.11), the racks holding the spool and the spindle are raised. This releases the pressure on the spindle (Figure 2.11). The lap spindle can then be partially withdrawn so as to take out the full spool. At the same time, the spindle is again passed through another empty spool. By again using the handle, the spindle holding the spool is lowered and the empty spool is made to rest on the fluted lap rollers. There

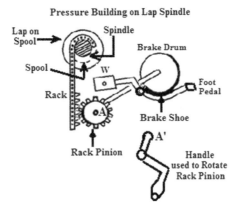

FIGURE 2.11 Pressure on lap spindle through racks:[3] The racks are responsible to make the lap with desired compactness. The brake shoes do not allow the racks to go up so easily unless the lap is sufficiently compact.

is an arrangement provided to level the spindle. This ensures perfect seating of the spool on the lap rollers with even and uniform contact and pressure across its width.

2.1.6.2   Ribbon Lap Machine[1,3]

This machine had been always coupled with the Sliver Lap machine to work as a pair. The object of this machine is to finally prepare suitable laps for the comber from the feed material (a lap) supplied by the Sliver Lap machine. The laps prepared on the Sliver Lap machine, have some shortcomings. The constituting slivers in the sliver lap do not get fully merged into the lap. The fibre parallelization is inadequate; although, 22 doublings of slivers forming a lap at the Sliver Lap try to compensate for the irregularity in constituent slivers along the lap length. However, for a comber lap, its regularity and uniformity across its width are as important as its regularity along its length. This is because; the same comber lap is required to be gripped across its width during combing by nippers (Figure 2.12). If the thickness of the lap across its width varies, the gripping of the lap during combing would not be uniform. The combing operation, in this case, would be seriously impaired. Uneven gripping during combing would also lead to the loss of good fibres (see ch. 7.12).

As shown in Figure 2.14, the machine has six heads and hence six laps made at the Sliver Lap are placed on fluted wooden lap rollers. The flutes hold the lap surface firmly and ensure the unwinding of the lap at the desired rate. Thus, for six heads, there are six drafting systems and all the laps are passed on to corresponding four-over-four drafting systems placed in front of corresponding heads. With a draft of around six, each lap is finally delivered by a corresponding drafting system onto six different curved plates (Figure 2.13). The draft of six is only arbitrary and a small variation, depending upon the final desired lap weight could always be made. As the drafting rollers carry out a similar function and controls a similar mass of cotton as in the Sliver Lap, the magnitude of the roller pressure is almost the same as applied to the drafting rollers of the Sliver Lap machine.

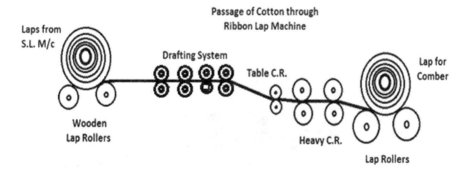

FIGURE 2.12   Passage of lap cotton through Ribbon Lap Machine:[1,3] The machine was basically used to improve the uniformity of the lap prepared at the earlier stage. However, the drafting system in older versions was the main source for not being able to control fibre movement.

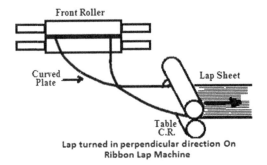

Lap turned in perpendicular direction On
Ribbon Lap Machine

FIGURE 2.13 Delivery of drafted sheet:[1,3] The direction of the material coming through the drafting system and that leads to superimposition of the drafted sheets are perpendicular to each other. The curved sheet is responsible to change this direction.

The curved plate on which thinned out drafted lap sheet is led is very smooth. Further it is so curved as to change the direction of delivery at right angles to that delivered by drafting rollers. The drafted sheets from the other drafting heads (Figure 2.14) are similarly passed on to the corresponding curved plates and then turned at right angles. The thinned-out lap sheet from each plate is guided on a smooth extended table with the help of corresponding pairs of table calender rollers. Thus, six sheets are superimposed on their way along the table so as to improve upon the lap regularity across the width and along its length. Finally, a compact lap comprising six layers is once again heavily calendered and wound on the wooden spools to make a comber lap.

The machine is provided with an arrangement for changing the draft in the drafting systems. The lap width is also selected so as to match with that required at the comber. Usually, this width varies from 10.5 to 11 in. A pair of Sliver Lap and Ribbon Lap machines normally feeds 6–7 Nasmith combers. As mentioned earlier, this old and conventional lap system almost lost its popularity with the introduction

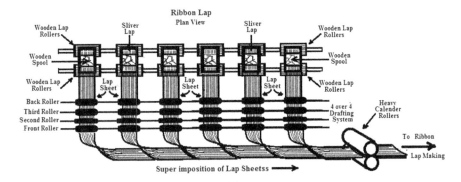

FIGURE 2.14 Ribbon Lap (plan view): A machine exhibiting a pioneering principle of superimposition of lap sheets.

of modern high-speed combers. However, in some mills where old and slow-speed combers were still working under the same roof, the system then continued to meet the requirement of conventional combers. In such cases, the system of Sliver Lap–Ribbon Lap was worked to produce much lighter laps (less than 600 grains/yard or 40 g/meter) for processing fine and super fine mixings only.

Even then, it needs to be mentioned that the concept of doubling a thin, drafted web sheet from each head has its potential advantage in improving lap uniformity both along and across its width.

2.1.6.3   Limitations of Conventional Sliver Lap–Ribbon Lap Sequence[3]

In the past, this sequence enjoyed great popularity and continued to hold its place with Nasmith combers. In the Sliver Lap (S.L.), there were 20–22 doublings of slivers; whereas in the Ribbon Lap (R.L.), six laps were doubled. This gave a total doubling of 120–132 in the pre-combing stage. All this was considered to be adequate in the past where the subsequent post-combing sequence was used to provide for doubling at speed frame and ring frame to compensate for any irregularity. The research work carried out, however, proved that a minimum level of pre-comb draft and number of doublings is necessary to get a satisfactory lap for the combing operation. The maximum total draft given in the Sliver Lap–Ribbon Lap sequence used to be around 12 (two in S.L. and six in R.L.).

Morton and Yen[4] showed that even after using this conventional lap preparation, about 40% of fibres were still hooked, though the preparation did help to increase the percentage of straight fibres from 25% to 50% (improved straightening and orientation). A certain minimum level of pre-comb draft and number of doublings were thus necessary to improve both fibre orientation and fibre extent. In the conventional sequence of S.L.–R.L., both the total draft and number of doublings were much less than this required level. It was also thought that the conventional Ribbon Lap machine was not able to sufficiently grip the heavy lap sheet, especially to give a draft of six. A typical defect called 'roller slip' was the reason for this.

In the Sliver Lap, owing to more bulk entering the drafting system, a high level of inter-fibre cohesion was experienced. To overcome this, the tension draft in each zone not exceeding 1.1 was used as a possible solution (easy drafting). However, this very much limited the total draft capacity of the sequence, restricting it to not more than 1.35. Alternately, it was thought to defunct the whole drafting system and to use the machine for merely assembling the slivers to form a lap. However, this imposed an extra burden on the Ribbon Lap machine to control both the heavier lap and higher draft.

Though the improved versions of the Sliver Lap and Ribbon Lap machines made a lot of improvements in the drafting system and production rates, the machine sequence still falls short of the required level of draft and doublings as compared to the latest lap preparation systems. It would be, therefore, worthwhile to compare the performances of laps made on a conventional system and a modern lap preparation system.

In a few mills, the Ribbon Lap machine was discarded. However, the Sliver Lap machine was retained and was used as merely a lap-making machine. This almost

simulated Platt's typical sequence of making the comber laps with the Draw Frame–
lap former sequence. Even then, this sort of improvization was good enough only for
finer mixings where the lap weights were not very high.

2.1.7 Improvements in Sliver Lap Machine

The new model introduced by Rieters has incorporated many improvements in
machine design. The conventional four-over-four drafting system is replaced by
a four-over-six drafting system. The spring weighting or the pneumatic weighting along with improved drafting makes it possible to have more positive control
over the fibres during drafting. This improves fibre control even at higher lap
weights. Pressure on the drafting rollers of up to 90 kg is possible. The calendar
rollers are also heavily weighted by pneumatical pressure to give a more compact
lap. This allows heavier laps to be produced and processed with more compact
lap sheets having almost no lap licking. The full-lap doffing is entirely auto-
pneumatic. It holds the lap body and then raises it for doffing. The empty spool
replaces the full lap. It is then repositioned on the lap rollers. All this operation is
quick and hardly takes less than 10 seconds. It also ensures a high level of safety
during the lap change making the operation quite simple without involving any
human assistance.

The stop motions have been coupled with a signal-lamps system. The sliver break
at the creel, roller lapping and full doff – all are indicated by the signalling system. On all the high-speed components, high-precision anti-friction bearings are
provided; whereas on slow-moving parts, plastic bushes do the same job. The automatic oiling system with oil pumps ensures good lubrication which further leads to
reduced maintenance and long service life with high reliability. It is thus possible to
increase operating speeds up to 60 m/min (200 ft/min) to bring about a substantial
increase in the output. Thus, the improvement in the quality of laps is brought about
at a considerably lower production cost.

2.1.8 Improvements in Ribbon Lap Machine

The new model introduced by Rieters has a three-over-four drafting system
(Figure 2.15). The nipping distance in the main drafting zone is adjustable from 30
mm to 54 mm, while that in the back zone can be varied from 32 mm to 56 mm. The
total draft in the machine can be varied from 4 to 9. The fluted bottom rollers and
synthetic covered top rollers are made with high-precision, anti-friction bearings.

The slow-moving parts such as the feed equipment and backline rollers are fitted with Teflon bushes which require little maintenance. On the older Ribbon Lap
machine, the loading on the drafting system was very much inadequate as compared
to the heavy weight of the material that was processed.

In modern sequence, a wide range of drafts and much higher loading on the drafting rollers are possible. This is certainly helpful when the laps of varying weight
per unit length are processed. The most modern and advanced drafting system on
the Sliver Lap and Ribbon Lap (S.L.–R.L.) offers excellent fibre control and gives

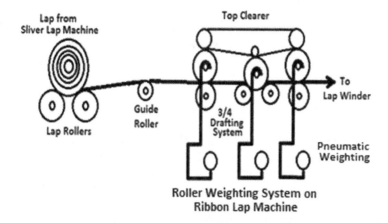

Roller Weighting System on
Ribbon Lap Machine

FIGURE 2.15 Pneumatic weighting on rollers:[3] Very high pressure of the air enables precise loading on the drafting rollers and this prevents defects such as roller slip.

ideal lap drafting conditions (Figure 2.16). Very high lap weights up to 75 g/m (1,060 grains/yard) with production speeds up to 60 m/min are possible.

The signal-lamp system provided on the new sequence is very efficient. It helps in better supervision and easy location of the causes of machine stoppages. A yellow lamp indicating the machine is in running condition, a red lamp pointing out lapping around the calendar roller or any other drafting roller, a green lamp showing the machine is ready for doffing (with manual doffing) – all help in improving the operating efficiency. The switches for starting or stopping the machine are very conveniently located so the operatives to reach them easily from all sides of the machine

The safety switches ensure that the doors for the headstock and hood are fully closed before starting the machine. This minimizes the risk of accidents. Inching motions* are provided for piecing-up operations. The pressure variation on drafting rollers is sensed by the control switches so that at any time, when the pressure falls below a certain level, the machine is automatically stopped. This avoids any deterioration in the quality of the material drafted.

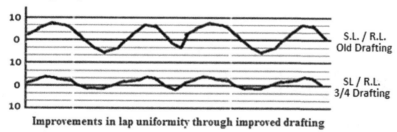

Improvements in lap uniformity through improved drafting

FIGURE 2.16 Lap irregularity with old and new drafting:[2,3] A new drafting system provided both the higher pressure required to handle higher bulk (lap weight) and improved fibre guidance. This enhanced the lap uniformity.

---

* With the button pushed or released, the machine starts or stops respectively.

The lap, before winding on the spool is very heavily calendered so as to make it compact. The full-lap weight up to 20 kg is possible and the laps when doffed are automatically weighed so as to indicate variations. An earlier version had manual doffing whereas; the automatic doffing using pneumatic pressure is the striking feature of later modern versions.

2.1.9 LAP FORMER[3]

Platt's lap former is a special machine resembling the combined features of both the Sliver Lap and the Ribbon Lap. It has a sliver-feeding system similar to the one on the Sliver Lap. However, as many as 40 slivers can be fed to this machine (Figure 2.17 and 2.18). The slivers after passing through a pair of top and bottom sensing rollers (electrical connections given to both for sensing the sliver break) enter 2/3 drafting systems equipped with top arm spring weighting.

The machine provides a maximum draft of up to four. There are four separate drafting heads and each has its separate creel at the back. Each head takes ten slivers and employs a draft of around four for the group of ten slivers. The drafted thin sheets emerge from each head in the front (Figure 2.18). These four sheets are turned at a right angle over smooth curved steel plates.

The sheets during their journey ahead are calendered and superimposed. Finally, a single combined sheet of the lap of the required width, thickness and weight suitable for the comber is again heavily calendered and wound in the roll form on the spool. It is believed that the superimposition of four sheets improves the uniformity of the lap both widthwise and lengthwise. The normal lap width is around 10.5 in.

In modern versions of the lap former, pneumatic loading is used to compress and consolidate the lap sheet. Thus, it is possible to produce heavier laps maintaining the same compactness. The machine is equipped with a six volt supply for stop motion circuits and the warning signal lamps at each delivery are provided to indicate roller lapping, sliver break, full doff, etc.

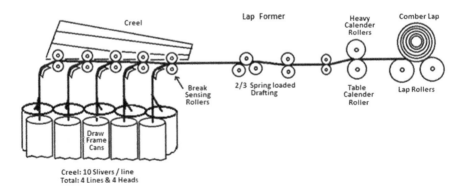

FIGURE 2.17 Passage of cotton through lap former:[1,3] Improved drafting system and increased number of doublings on lap former produced laps with better uniformity both along the length and across it.

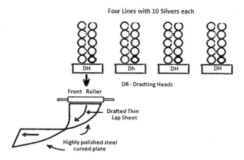

FIGURE 2.18   Creeling on lap former:[1,3] In all 40 slivers are creeled and arranged behind the drafting system and they are divided into groups of ten each. Here too, the use of the curved plates enables the change in the direction of the material through 90°.

2.1.9.1 Drafting System

A typical two-over-three drafting system (Figure 2.19) is a single zone drafting and the second top roller runs on both the second and third bottom roller.

This improves the grip on the heavy lap. The setting between the second and third bottom roller is thus fixed (1.75 in); whereas that between the front and second bottom roller can be changed to suit the staple length of the cotton being processed. As claimed by the manufacturers, this setting is usually E.L. + 3 mm (0.125 in). This again varies a little between 0.0625 to 0.125 in (1.6 to 3.0 mm). The lower setting is used for higher drafts up to 4.0 in; however, the machine is often worked with lower drafts below 3.0 in.

2.1.10   SUPER LAP FORMER[1]

With the advent of high production comber, the demand for better lap preparation increased. This was necessary because the comber was designed to process heavier laps (one of the features of high production). This necessitated better lap preparation

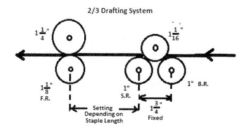

FIGURE 2.19   Drafting zone:[1,3] A special single zone drafting system to minimize the phasing effect due to piecing wave. The bigger diameter rollers permit very high loading on top rollers,

taking due care of both lap uniformity and fibre parallelization. Therefore, with higher lap weight, it became essential to improve both the fibre orientation and fibre disentanglement. This again was necessary because it required more effective penetration of cylinder needles without causing any damage to fibres or needles. The better lap preparation also enabled the comber to do better fractionation. The practical limitation, however, arises when higher pre-comb drafts make the fibres in the lap more parallel, thus leading to a typical problem – 'lap licking'. This leads to more soft waste. Apart from this, the uneven feeding owing to lap licking leads to uneven sliver at the comber. The operator minding the comber machine has to be careful in timely noticing the licking phenomenon and removing the thicker portion. It is here that it leads to more soft waste.

Whitin and Sons overcame this difficulty by increasing the draft on the first pre-comb machine (drawing frame) and reducing its value on the final super lap former.

The typical sequence thus consists of the Whitin Draw Frame and Super Lap Former (Figure 2.20). The Whitin Draw Frame has a very precise and robust four-over-five drafting system with a maximum drafting capacity of ten. In normal working, however, the draft is around eight, with eight sliver doubling.

The advantage of using draw frame drafting first in the pre-comb sequence is twofold: first, the material is comparatively lighter in weight (individual sliver weight), when the slivers are drafted. This gives better control over the fibres during drafting to improve its uniformity even at higher production rates and secondly, the workload on the final lap-making machine – the Super Lap Former – is reduced. This allows lower drafts at lap-making process, thus reducing lap-licking problems.

There are three sliver creels with stands. Each creel holds 20 slivers (Figure 2.21 and 2.22). The creel is double-sided and holds 10 slivers each on either side.

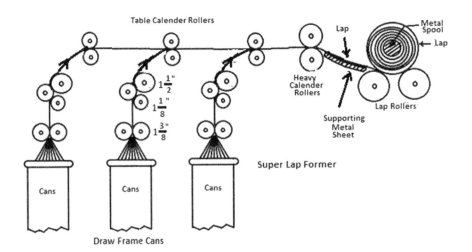

FIGURE 2.20 Passage of cotton slivers through Super Lap Former:[1,3] The concept introduced by Whitin and Sons to make compact and heavier laps for their then high-speed combers.

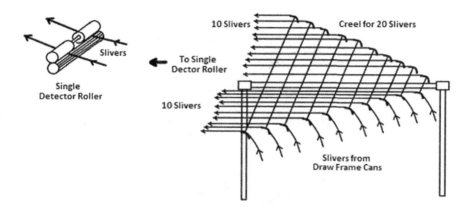

FIGURE 2.21 Super lap creel:[1,3] Each drafting head takes up 20 slivers, which are further divided into groups of ten each. The creel is specially designed to accommodate these two groups which again are merged into a single lap sheet.

Using a double-sided creel also shortens the length of the creel extending behind the machine. Thus, with 3 creels at the back, as many as 60 slivers can be fed into the machine. The sliver guides suitably placed above each sliver-can, are attached to the three creels. This allows straight vertical withdrawal of slivers without touching the periphery of the can and reducing any possibility of a false draft. As usual, the slivers are lifted by the pairs of rollers and are guided on the creel table. The rollers while lifting the slivers from the creel-cans also sense any sliver break.

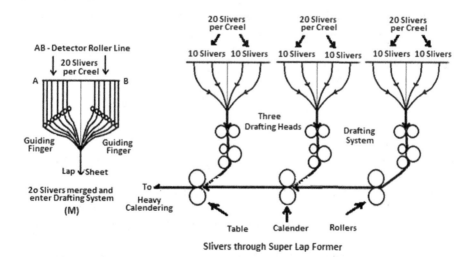

FIGURE 2.22 Sliver coming from creel of Super Lap Former:[1,3] In each creel head, 20 slivers in the group of 10 each are superimposed in such a way that the sliver lap uniformity across the width of the lap is improved.

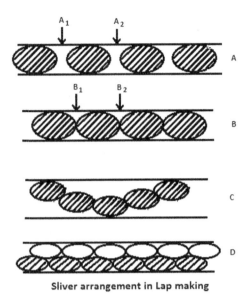

Sliver arrangement in Lap making

FIGURE 2.23 Sliver arrangement in lap:[1,6] Out of the four different arrangements, only 'D; gives the best lap uniformity across its width; because the lap thickness across the lap width is far more uniform in this arrangement.

The slivers, in a group of ten (from each side of the creel) then travel vertically downwards to enter the two-over-three drafting system (Figure 2.22). Thus, for each creel, there is a separate drafting unit and 20 slivers are all accommodated in one drafting system. The lap sheet processed through each drafting system occupies the desired width suitable for a comber lap.

Ordinarily, when the slivers are laid side-by-side to form a sheet of a lap, as they enter the drafting, it is necessary that there should not be any gaps between the adjoining slivers. As seen from Figure 2.23, their loose positioning (wider gap) would form a lap with intermittent thick and thin places.

As shown (Figure 2.23 A), the lap thickness at A_1, A_2 – would be forming necks (thin place). The variation is bound to occur even when the slivers are closed down (Figure 2.23 B). The necks are bound to form at B_1, and B_2. The role of sliver sheet guides, placed just before the drafting system, have to take care of the correct positioning of slivers to avoid any overriding (Figure 2.23 C) of slivers at the sides.

In the Super Lap, the slivers enter the drafting system with a formation as shown in (Figure 2.23 D), and this arrangement significantly improves the uniformity of lap thickness. The vertically arranged drafting system (Figure 2.24) has a pair of back rollers in front of which there is a one-over-two front drafting roller system. The second bottom roller is smaller in diameter to permit a closer setting for short-staple cotton. The top rollers, as usual, are covered with synthetic rubber and are loaded with spring-loaded top arms. A very heavy pressure of 110 kg (250 lbs) is put on each top roller. The webs as drafted through the three corresponding drafting systems come

2/3 Drafting Particulars

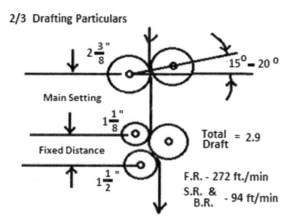

FIGURE 2.24 Setting of drafting zone:[3,7]: As the single top roller runs over the first and second bottom roller, their distance is fixed. Thus, it becomes a single-zone drafting system. The second top roller overhang in the backward direction provides better fibre guidance.

out and are lightly calendered. The calender rollers are positioned over an extended, smooth, polished steel table. On this table, the three webs are suitably superimposed, again very heavily calendered with a pressure of 360 kg (800 lbs) and finally wound as one compact sheet on the steel spool.

Among the three drafting heads, it is possible to slightly vary the draft independently. Figure 2.25). As the lap licking is influenced by the amount of draft employed, the independent control of draft in the three drafting heads is a distinct advantage. The average draft employed in the three drafting systems can be expressed as:

$$\sqrt[3]{\left(D_1 \times D_2 \times D_3\right)} \text{ – where } D_1, D_2 \text{ and } D_3 \text{ are individual Drafts}$$

Further, the licking is a surface phenomenon and thus it is possible to reduce the draft levels in two extreme sheets forming the top and bottom layer, while the draft

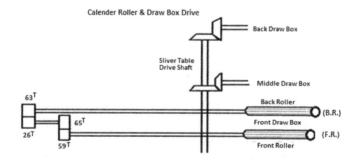

FIGURE 2.25 Changing draft in individual drafting head:[1,3] This provision enables adjusting the draft in the outer lap layers. It helps in reducing lap-licking tendency.

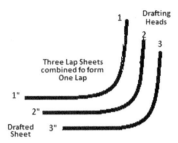

FIGURE 2.26 Draft level adjustments:[1,3] Normally the draft for the outer layers is proportionately reduced, whereas the draft for the inner layer can be slightly increased so as to maintain the desired lap weight.

for the middle layer can be slightly increased. It, however, needs to be investigated whether the fibre orientation in the three layers with the above arrangement significantly differs.

The maximum possible draft can be kept in drafting head 2 (Figure 2.26) which lies in between the two outer layers coming from heads 1 and 3. The draft in the other two heads can then be reduced so as to have the same average draft. It can be seen from the figure that layer 2 with more draft nicely gets sandwiched and does not come in contact with the outer lap sheet surface. Whitin the Super Lap Former is self-content for its lap doffing operations and embodies positive acting mechanisms. These are activated by integrated pneumatic and electrical controls.

A lap winder head thus completes the following operations automatically in sequence:

1. Winding the lap on a spool.
2. Stopping the machine when the lap is full and simultaneously cutting the lap sheet.
3. Raising the spool from winding drums.
4. Widening the lap flanges to release the spool. The spool is then allowed to roll down the tray for the operative to pick it up. The operative, however, puts it manually on the trolley carrying the laps to comber.
5. Replacing the empty spool in its position.
6. Securing the lap flanges to hold the spool and then lowering it down to correctly position it on the winding drum.
7. Restarting the machine for winding a new lap.

All this takes place within 10 seconds. With the machine speed of 245 ft/min (74.3 m/min – conventional machine), each lap weighing approximately 13–14 kg, the time taken for completing one full lap is less than 3 minutes. Earlier versions had a manual doffing system, whereas the later version was equipped with fully auto doffing. On average, one unit of Draw Frame and Super Lap then fed approximately eight combers.

When the full-lap spools are in plenty or else empty spools are not available, there is a provision to run the machine manually as and when required. Sometimes, there was a tendency in the mills to run the machine with half-full laps, when there was a shortage of laps at comber. Rather than taking short laps, which may be a short-term measure; it is much better to correctly balance the production of super lap former with that of comber.

The opening and closing of lap flanges to remove the full lap and to reload empty spools as well as raising and lowering of the slide frame holding the lap fingers at the time of doffing – are all done pneumatically. For this, the compressor is required to supply clean, filtered air. The pneumatic pressure ranging from 45–50 psi (3–3.5 kg/cm^2) is necessary for the operation. A full-limit switch or length counter stops the machine when the lap is full and the signal is passed to the doffing mechanism. Defects such as lap flanges closing or opening too fast, flanges being forced outwardly during lap building or soft laps – all are due to either the wrong setting of the control valve or wrong pressure. The faulty pressure is sometimes due to airline leakage.

The design of the Super Lap Former fits very well with modern trends and requirements. The stop motions are provided for sliver breaks at the creel and roller lap-ups in the drafting zone. The light indicators glow appropriate coloured lamps for showing the cause of machine stoppages. The machine is capable of giving a maximum total draft of 3.65 with 60 doublings. Heavy laps of up to 1,200 grains per yard (75 g/m) can be produced. The heavy lap weight alone helps in giving almost double production rates at the comber. The production of the Super Lap itself can be in the range of 260 to 300 kg/hour.

2.1.11　Effect of Lap Preparation Systems on Waste Extracted at Comber[1,3]

It is already known that the number of processes between card and comber has to be even to take the advantage of presenting the majority of the hooks in the leading direction. It has been also established that with reference to the condition of the fibre disorder in a card, a certain amount of pre-comb draft is essential to parallelize the fibres sufficiently so as to remove much of the fibre entanglement and improve the fibre extent. Thus, a well-prepared lap with due attention to both these factors stated above helps to reduce:

1. The waste is extracted at the comber.
2. The strain on cylinder needles.
3. Fibre damage.
4. Chances of good fibres going to waste.
5. Short fibres going into the sliver.

The commonly known lap sequences in the mills are – (a) Sliver Lap–Ribbon Lap (S.L.–R.L.), (b) Draw Frame–Lap Former (D.F.–L.F.), (c) Drawing Frame–Sliver Lap (D.F.–S.L. – optional) and (d) Drawing Frame–Super Lap Former (D.F.–S.L.F.).

As mentioned earlier, the S.L.–R.L. combination with a graduated drafting system causes serious irregularity owing to roller slip, especially owing to heavier sliver mass in the drafting zone. These irregularities appear as medium-term variations in the comber head sliver and long-term variations in the yarn. For this reason, D.F.–L.F. system is preferred to this conventional system. However, when it is necessary to use S.L.–R.L. system for some reasons, either the higher lap weights should be avoided or else, the higher lap weights should be reached by reducing the draft in the S.L. machine to 1.2 and also converting the R.L. drafting from four-over-four to three-over-four (two zones) drafting with a total draft ranging from 3.5 to 4.5. In this case, the back zone draft (in R,L,) would range from 1.4 to 1.7.

With D.F.–L.F. system, the number of doublings is 320 (40 in L.F. and 8 in D.F.). The system provides comparatively better parallelization of fibres in comparison to S.L.–R.L. sequence (total draft in D.F.–L.F. is 20–24, as against 12 in S.L.–R.L.), and it is capable of producing more uniform and heavier laps. There are some versions of D.F.–L.F. which have been tried for academic interest.

D.F.–D.F.–L.F.: An additional passage of the drawing frame was also tried in the pre-comb stage. Thus, the total pre-comb draft and number of doublings were increased.

D.F.–D.F.–D.F.–L.F.: Some researchers also tried two additional drawing frame passages so as to maintain an even number of passages between card and comber.

In the first sequence, the total draft and number of doublings are 36 and 1,440 respectively (assuming very little draft in L.F.); whereas in the second sequence, the corresponding values are 216 and 8,640. Both these sequences gave a much higher draft and a very high number of doublings. However, in the first, the direction of presentation of fibre hooks was unfavourable; in that, the majority of the hooks were presented to the comber in a trailing direction. In the second, the direction of hooks was favourable; however, the amount of draft was much higher than that required for the level of desired parallelization. In fact, the higher draft in the second sequence led to excessive parallelization thus resulting in lap licking at the comber.

It is proved that even after six to eight passages of the draw frame; the effect of fibre parallelization is still detectable. But this effect becomes relatively much smaller beyond a draft level of 36. It is observed by many researchers that beyond a draft level of 30, the difference, though noticeable, is not very significant. Hence, it is very appropriate to restrict the draft up to 30 in the pre-combing stage.

This gives substantial savings in the waste extracted at the comber. It may also be noted that the yarn quality in terms of yarn tenacity and C.V.% at this level of the draft, is not significantly different than that at higher levels. Restricting the draft up to 30 also helps in promoting the 'self-cleaning' aspect discussed earlier. (Section 2.2.5). Further, it needs to be pointed out that an unnecessary higher number of passages in pre-comb preparation involve only additional machines, labour cost, space and more power.

Obviously, this additional cost has to be compensated by saving in the comber noil. When more than two passages are involved in pre-comb preparation, the equivalent saving in comber noil does not compensate for the increase in the cost due to more machines, labour, space, power, etc. With better lap preparation, involving

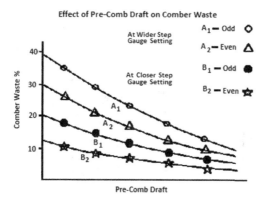

FIGURE 2.27   Effect of pre-comb draft on:[1,3] Increase in pre-comb draft improves fibre parallelization and orientation. This helps in reducing the comber noil without changing the settings. However, this is further influenced by the step gauge setting. A closer step gauge setting significantly reduces the comber noil.

only two machines in a pre-comb sequence, more benefits could be accrued. This is because; the better lap preparation, with two machines, is able to give fairly uniform and heavier laps and at the same time there is an appreciable reduction in the comber noil.

The lap preparation system like – Draw Frame–Super Lap Former (D.F.–S.L.F) met all these technological requirements. It involved a minimum number of machines in pre-combing stages, offered an optimum draft level of up to 30 and provided as many as 480 doublings which were more than adequate. In fact, no other sequence was able to meet the requirements so completely. Both the lap weights and the width were adjustable in this sequence and hence, with little modifications, it still can be adopted for any modern combing operation.

The graph (Figure 2.27) reveals that the difference between the odd and even number of machines is quite distinct at the lower pre-comb draft. However, with a progressive increase in this draft, the number of machines whether odd or even (direction of the majority of the hooks) does not matter much. This is because, the higher level of draft improves fibre orientation, reduces fibre entanglements and, to a great extent, straightens the hooks. The graph also reveals that the influence of the step gauge setting is more significant at lower pre-comb drafts. All four graphs (Figure 2.28) become closer as the pre-comb draft is steadily increased.

2.1.12   SOME EXPERIMENTS WITH PRE-COMB DRAFT AND FIBRE PRESENTATION[3]

It has been well stressed that the comber is very sensitive to the direction of fibre presentation. Earlier, some research work was carried out where the laps were prepared by a pre-comb machinery sequence involving two machines. The effect of the direction of presentation of the majority of hooks was examined by feeding the lap in the normal (majority of hooks in the leading direction) and reverse direction to the

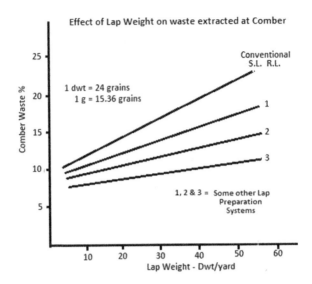

FIGURE 2.28   Effect of lap weight:[1,6] An increase in the lap weight, if it involves poor fibre parallelization and orientation, invariably leads to an increase in comber noil%.

comber. Thus, the majority of the fibre hooks presented in each case being different, it was expected that the comber would treat them differently.

It is interesting to note that with the increase in the pre-comb draft, the reduction in the noil with the majority of the hooks presented in trailing direction was 47.82% i.e., {[(32.0 − 16.6) / 32.2] × 100}; while that with the majority of leading hooks presented was 40.4% i.e., {[(24.0 − 14.3 / 24.0] × 100}(Table 2.1).

This clearly brings out the fact that the comber, though sensitive to the direction of presentation, is not so when sufficient parallelization level has been achieved. This is because; as the pre-comb draft is increased, the overall level of parallelization gets improved. The sensitivity of the comber can be seen when the pre-comb drafts are low. The table shows this difference, at 'zero' pre-comb draft, in the form of values of comber waste extracted as 32.0% with the majority of hooks in the trailing direction and 24% with the majority of hooks in the leading direction, the difference is 8% in the extracted waste. However, with the pre-comb draft being as high as 16, this

TABLE 2.1
Effect of Fibre Presentation on Comber Performance

	Percentage of Comber Waste Extracted with		Difference in Noil%	
Pre-Comb Draft	Majority of Hooks in Trailing direction [A]	Majority of Hooks in Leading Direction [B]	Absolute Difference	% on A
0	32.0	24.0	8.0	25.0
4	21.1	16.2	4.9	23.2
8	18.4	15.4	3.0	16.3
16	16.6	14.2	2.4	14.4

difference is comparatively much smaller (16.6 − 14.2 = 2.4%). It seems, therefore, that with higher draft levels, there is better parallelization and orientation of fibres which reduces the total hook content and increases fibre extent. Hence, the difference in the comber noil extracted with normal and reverse directions seems to have reduced and has been much less influential and conspicuous.

2.1.13   STAPLE LENGTH AND PRE-COMB DRAFT[1,3]

It has been proved that the degree of parallelization, for a given draft, depends upon the staple length of the fibre. Thus, short-staple cottons are more easily parallelized than longer ones. This obviously would influence the choice of lap preparation because; with a lesser draft given in pre-comb preparation, the required level of parallelization may be easily achieved when using short-staple cottons. Any additional passage or higher level of the draft would bring only marginal savings in comber noil. Under these conditions, it is found that a pre-comb draft of 12–15 for short-staple cotton would be quite adequate, provided that a correct direction of fibre presentation is maintained.

An interesting experiment was carried out at Smyre Manufacturing Company (North Carolina, United States) in 1958 with American cotton, where three laps were prepared – (1) D.F. – lap winder (2) D.F.–D.F. – lap winder and (3) D.F.–D.F.–D.F. – lap winder.

With first lap preparation, the comber noil was 13.75%; with the second method, it was 11.4%, though the yarn tenacity dropped by 4% in this case. However, there was no change in the yarn uniformity as measured by Uster. In the third trial, comber noil was 10%. Truly speaking the noil in the second case should have been more than that in the first case; because of the unfavourable direction of the majority of fibre hook presentation according to Morton.[7]

Against the longer staple, Egyptian cotton used by Morton in his experiment, in the above example, American cotton is used; it can be assumed that the level of parallelization achieved through pre-comb draft must be different in these two cases.

2.1.14   EFFECT OF HEAVIER LAPS ON COMBER NOIL[1,3]

The demand for heavier lap weight arose mainly because of higher production rates sought in the comber. When increasing combing speeds, some fundamental improvements in machine design were essential. Similarly, when increasing the lap weight some changes in design – nipper grip, nipper pressure, stronger needling, etc. were required. Only then, it directly enabled higher output. However, it is essential to keep the performance of the comber in terms of fractionation at the same level; otherwise, the advantage of increasing production would be more than offset. With higher lap weight, it is, therefore, necessary to improve both fibre parallelization. and orientation for which the pre-combing preparation must be improved. It was found that when the comber lap weight was progressively increased, keeping the same lap preparation, the waste extracted at the comber also increased.

This means that there is insufficient parallelization and orientation of the fibres in the lap. Especially, when the laps were prepared with a lower number of doublings and pre-comb draft, this increase in comber waste was more marked. It may be noted here that such an increase in the comber noil does not bring any improvement in the combing quality.

It is, therefore essential that, with heavier laps, the lap preparation needs to be improved. This saves some good fibres from unnecessarily going to waste. In one typical experiment, the laps of varying weights were prepared with a different sequence (Figure 2.28). It is clearly seen from the graph that when the lap weight is increased, the increase in the comber waste with poor lap preparatory sequence is much more marked. As compared to this, the increase in waste with other lap preparation, meeting the necessary requirements gave comparatively lower rise.

It can, therefore be seen that the old conventional S.L.–R.L. sequence with both inadequate draft and the number of doublings cannot cope-up with higher lap weights. Graphs 1, 2 and 3 (Figure 2.28) are for some other better lap preparation systems. Amongst the three other graphs, graph 3 seems to be the best. In this case, even when the weight of the lap is increased beyond 50 dwt/yard (more than 85 g/m; i.e., 1 dwt or one pennyweight = 1.555 g), the comber waste increases only marginally. It is again evident that the conventional Sliver Lap–Ribbon Lap sequence not only falls short of the amount of draft and number of doublings that are required, but also fails in handling the heavier laps, a pre-requisite for higher productivity. Thus, one of the equally important requirements for targeting higher production rates at the comber is supplying the suitably heavier comber laps that would invariably require a far better lap preparation system.

Hence, while making heavier laps, the number of doublings, pre-comb draft, the type of drafting system used in preparation and control over the heavier weight – all associated with any pre-comb sequence are important. This is because all of them decide the quality of the laps prepared.

REFERENCES

1. *Manual of Cotton Spinning: "Draw Frames, Comber & Speed Frames": Frank Charnley*, The Textile Institute Manchester, Butterworths, 1964
2. *Spun Yarn Technology: Eric Oxtoby: U.K.* Butterworth Publication, 1987
3. *Elements of Cotton Spinning: Combing: Dr. A.R.Khare*, Sai Publication
4. *Morton & Yen – J.T.I.* 1953
5. Morton & Summars, *Journal of Textile Institute*, 1949, Vol.40, p.106
6. A practical guide to Combing and Drawing – W.Klein, *Manual of Textile Technology*, The textile Institute Publication
7. Morton & Yen, Textile Institute, 1953, T-317

3 Comber

3.1 COMBER

3.1.1 INTRODUCTION

As mentioned earlier (Section 1.1.3), the prime object of the comber is to separate and remove the short fibres, thus upgrading the cotton. Again, apart from this, it also removes a large number of neps and foreign matter. But this is only incidental. Further, the control over the neps has to be more precise in carding rather than in combing. This is because; it would be more economical to run the card with better management of card wires (sometimes even at a lower production rate) to keep the nep level to a minimum rather than trying to extract more waste at the comber by merely removing the neps.

As the cotton fibre is much shorter in length as compared to many other fibres, the continuous method of combing is not possible. The combing operation, therefore, has to be carried out intermittently. The sequence of operation is as follows.

3.1.2 CYCLE OF OPERATION[2]

The laps from the combing preparatory sequence (final lap-making machine) are kept on the wooden lap rollers. These rollers (Figure 3.1a) are intermittently driven through a pawl and ratchet mechanism. In the figure, though only one head is shown, the number of heads for a machine varies from six to eight and hence there are as many laps as the heads.

The lap thus unrolled is guided over tension compensating plate the ABC (Figure 3.1b) which is hinged at (B). The point C, along with the nipper frame, moves forwards and backwards, as shown by the arrow in Figure 3.1c and hence in the forward-most position, the plate ABC almost straightens out (Figure 3.1b); whereas in its backward-most position, it folds around point B (Figure 3.1c).

The main function of this plate is to take-up the extra length of the lap between the feed roller and the lap roller when point C is in the most backward position (i.e., when the nippers are also in the most backward position. However, the same length is again made available when point C moves with nippers and is in the most forward position. Apart from this, the plate itself guides the sheet of the lap from the lap roller to the feed roller. The feed roller (Figure 3.2) is held and suitably mounted on the bracket held by the bottom nipper. The lap is received and guided through the feed roller and the bottom nipper plate. Like the lap roller, the feed roller too rotates intermittently and the timings of these two are perfectly synchronized.

Further, the length of the lap delivered by these two is also equal. Sometimes, instead of a single feed roller, there are two feed rollers in typical conventional

DOI: 10.1201/9780429486555-3

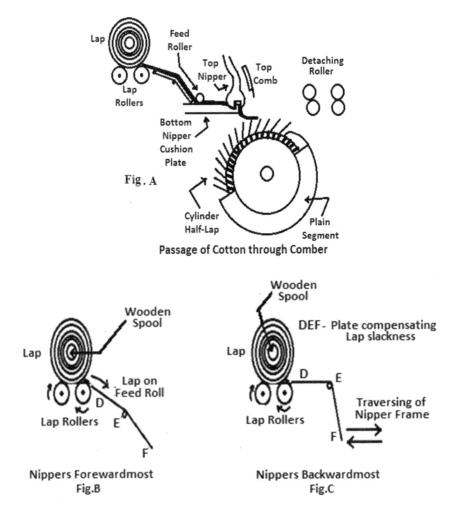

Passage of Cotton through Comber

Nippers Forewardmost
Fig.B

Nippers Backwardmost
Fig.C

FIGURE 3.1 (a) Passage of cotton through the comber:[1,2] The combing cycle is conveniently divided into two operations – (1) combing by cylinder and (2) piecing-up operation. The figure shows the first operation (b) Tension Compensating Plate:[1,2]While compensating for tension, it does not allow the lap to fold during the combing cycle.

machines and they too carry out the same job. The lap is then received by a pair of nippers – the top and the bottom (Figure 3.3). Both are specially shaped to perfectly grip the fleece issued by the feed roller. This gripping is very much essential during the combing by the cylinder needles. The bottom nipper is positioned securely whereas, the top nipper is movable.

When the lap is required to be gripped, it is the top nipper which closes down onto the bottom nipper. When the operation of combing by cylinder needles is over, the top nipper is lifted (opened out) and thus the lap is made free. It is, however, essential that during combing, the top nipper sits perfectly on the bottom nipper, with a lap in between, so that no fibres thus gripped are taken away by the cylinder needles.

When the nippers are fully closed, a part of the lap fringe projects in front of the nippers (Figure 3.4). Actually, this fringe is subjected to the action of fine rows of cylinder needles. These needle rows are, in fact, a whole set of needle strips, which

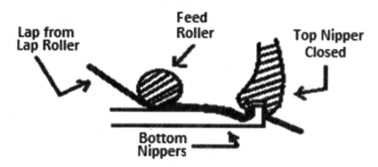

FIGURE 3.2   Feed rollers:[1,2] The main function of the feed roller is to deliver a pre-determined length of the lap in each combing cycle.

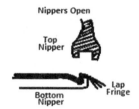

FIGURE 3.3   Nippers open:[1,2] In each combing cycle, the nippers remain open for a certain time to allow the lap to be pushed ahead.

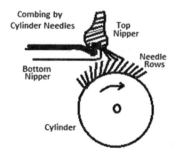

FIGURE 3.4   Lap Fringe presented to needles:[1,2] During the main combing operation, the tail end of the lap fringe is firmly held between the top and bottom nipper.

are securely mounted on a segment called 'half lap'. The half lap can be detached from the bare cylinder for any maintenance work. Especially, when the needles are required to be freshly mounted on a needle strip after soldering is required to be re-fixed on the half lap. The half lap is screwed to the inside bare cylinder and thus, its mounting or dismounting becomes an easy operation. In old combers there were 20 rows of needles on a half lap; whereas, with modern combers, there are only 17 rows.

These rows of needles pass through the lap fringe projecting in front of the nipper grip, one after the other, and remove anything which is not gripped by the nippers. Usually, the shorter fibres have less fibre extent and as such are not gripped by the nipper bite. The cylinder needles, therefore, can easily remove from the fringe. It is equally possible that the fibres having trailing hooks are also judged by the needles as short fibres.

Particularly, with the trailing hooks which reduce the fibre extent, if the tailing ends of the fibres are out of the bite of the nippers at the time of combing by cylinder, they are also removed by the needles. As for the fibres with leading hooks, the needles conveniently pass through the hooked portion and try to straighten them out, thus increasing their fibre extent. After the cylinder combing operation is over, the cylinder in its course of rotation brings the plain segment (Figure 3.5) to its upper side. During the same time, the nippers advance ahead and also open out during this journey. The fringe which has been combed by the needles is brought towards detaching rollers. Around this time, the detaching rollers are turned back (rotated in the reverse direction) for a short duration. This results in bringing back some portion of the fringe that is already combed in the previous cycle. The nippers in their most forward position reach closest to the detaching rollers.

The two lap fringes – one already combed and the other freshly combed are thus partly superimposed (Figure 3.6). This joining of the two fringes is called a 'piecing-up operation'. Thereafter, two things follow immediately and also simultaneously. The first is that the detaching rollers start moving in their normal direction, taking up the joined fringe ahead.

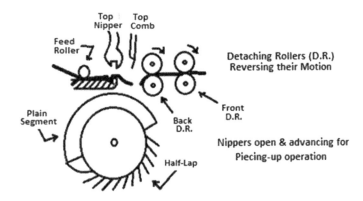

FIGURE 3.5 Piecing-up operation:[1,2] Nippers advancing to present the combed lap fringe to detaching rollers.

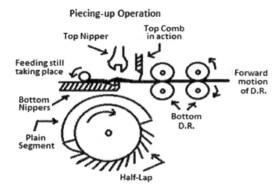

FIGURE 3.6 Piecing-up operation:[1,2] The top comb is withheld till the piecing joint of old and new lap fringe comes under detaching roller nip.

Secondly, when the joint of two fringes advances further and comes under the nip of the detaching roller, another comb, called the 'top comb' with a single row of needles, enters the fringe that extends behind the detaching rollers (Figure 3.7).

A little later, the feed roller starts advancing (issuing) a certain length of lap. The top comb merely drops down into the fringe and remains stationary in that position. Therefore, as the detaching rollers continue to pull the fringe, the needles of the top comb placed in the path help in combing the tailing portion of the fringe (Figure 3.7)). The influence of the top comb needles is such that, when it enters the fringe, those fibres which are not positively drawn ahead by detaching rollers, are restrained back and prevented from going ahead.

At this juncture, the detachment of the fringe really begins. All such fibres which are thus restrained gather behind or around the top comb needles. While the top comb is being lifted-up later, the nippers start moving back and also close. The next fringe is again firmly gripped and thus cylinder combing cycle begins again.

A rotating brush with fine and soft bristles is placed under the cylinder. The bristles of the brush are made to penetrate a little (Figure 3.8) into cylinder needles, thus enabling easy stripping of fibres and foreign matter embedded into needles. Rieter has developed special brushed where the bristles are flexible and yet stron. They withstand the normal friction against the strong 'Uni-comb' type of needles.[8]

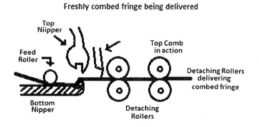

FIGURE 3.7 Delivery of the combed web:[1,2] The resultant motion of the detaching rollers in the forward direction ultimately delivers the lap fringe ahead.

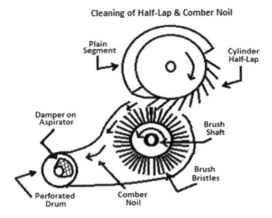

Cleaning of Half-Lap & Comber Noil

FIGURE 3.8 Half-lap cleaning:[1,2] In every cycle, as the needles remove short fibres and other foreign matter, they need to be cleaned. The revolving brush with its bristles do this.

The needles are thus cleaned and become ready for the fresh combing action in the next combing cycle. An exhaust created by a suction fan which is placed suitably underneath, is directed through a perforated drum and is made to reach very close to the surface of the brush. This helps in removing the fibres and foreign matter from the brush bristles to clean the bristles for the next combing cycle.

3.1.3 ASPIRATOR[2]

The mechanism which generates suction and directs it to the brush bristles is called an 'aspirator'. It contains a hollow perforated drum with a damper fixed inside.

As shown in Figure 3.9, the dampers have openings (A_1, A_2, A_3 and B_1, B_2, B_3). These openings are directed towards the brushes of the respective heads. The connections are made from the suction fan to the hollow damper through a suitable duct which is situated at the centre of the perforated, hollow drum.

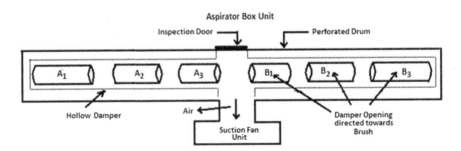

FIGURE 3.9 Damper in the aspirator:[1,2] The fibrous matter accumulated on brush bristles is ultimately removed by suction created inside the damper. The perforations on the drum surface allow the air to pass through whereas the fluffy short fibres with other impurities are retained on the drum surface.

The suction fan draws the air through the damper openings. This creates an air breeze which comes from around the brush bristles. This is how the noil adhering to the bristles gets stripped off. The outer perforated drum provides the way (inlet) for air coming from the brush. There are six openings for the damper and the suction is provided at the centre. It is, however, necessary to see that the strength of suction at each of these openings is adequate enough to create an air breeze around the bristles of corresponding brushes to strip them effectively. To ensure this, the openings – A_1 to A_3 and B_1 to B_3 are made progressively larger. In short, $A_3 > A_2 > A_1$ or $B_3 > B_2 > B_1$.

The fibres thus removed from the brush bristles and which ultimately get deposited on the outer perforated, hollow drum are in the form of a comparatively thin fleece. This fleece is the comber noil or the comber waste. It mainly consists of short and immature fibres, some foreign matter and neps. The deposition of waste on the perforated drum is again intermittent. A mechanism is provided to rotate the outer perforated drum intermittently. This helps in bringing a fresh and clean perforated surface to the oncoming noil stripped from the brush in each cycle.

It may also be mentioned here that when the outer drum rotates, the position of the inner damper with its corresponding openings remains unaltered and thus the suction is always directed towards the brush.

The fleece of waste accumulated on the perforated drum is then consolidated by a light dead-weight roller (Figure 3.10) so that a more compact fleece is formed. Finally, the waste fleece gets deposited into the waste tray suitably placed at the back and underneath the perforated drums.

In old combers during combing operation, the back top detaching roller was made to swing or rock back a little towards the cylinder (Figure 3.12). The rocking of the top detaching roller was around its corresponding bottom detaching roller. This made the already combed fringe keep itself aligned on the circumference of the bottom detaching roller and get ready for joining with freshly combed fringe. In fact, this joining of old and new fringe is called the 'piecing-up operation'. In all

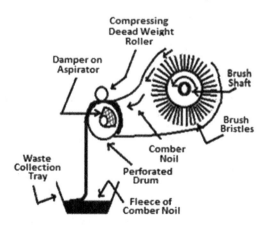

FIGURE 3.10 Collection and deposition of comber waste:[1,2] The deposited fluffy mass of short fibres and other foreign matter is deposited in the form of the layer in the trays.

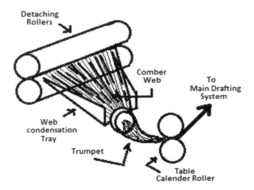

FIGURE 3.11   Web condensation:[1,2] The net forward movement of detaching rollers results in the delivery of a precise length of sliver web in every combing cycle.

modern combers, however, the rocking mechanism of the back top detaching roller is dispensed with. The web thus delivered by the detaching roller passes through the trumpet (Figure 3.11) and gets consolidated. It is then laid over a long smooth table and passed on to the main drafting system.

3.1.4   Detachment[2]

As explained earlier, the two fringes – already combed (in the previous cycle) and freshly combed are joined together. In fact, a little overriding of the two fringe ends (superimposition) is purposely allowed. This helps in making a better joint of the two fringes. The detaching rollers are then made to rotate in their normal direction. This delivers the joined combed fringe ahead in the form of a thin web.

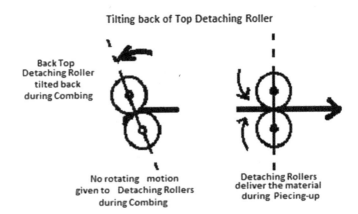

FIGURE 3.12   Tilting of back top detaching roller:[1,2] Though detaching rollers are stationary during cylinder combing operation, they tilt backwards to lay down the backend of already combed fringe.

The two pairs of detaching rollers have three types of motion, (a) they are stationary (i.e., do not rotate) during combing by cylinder, (b) they rotate a little in the backward direction for bringing back the already combed fringe for piecing-up operation and (c) they rotate in the forward direction, immediately after the piecing-up operation to deliver the fringe ahead along with the joint. It may be noted that the backward rotation of the detaching rollers is much less than its forward rotation. The difference between these two (backward motion and forward motion) ultimately corresponds to the resultant forward rotation. The corresponding equivalent length is called 'net delivered length'. This takes place in each combing cycle.

For a comber, the length of fringe reversed in a backward direction and that is delivered in a forward direction are all pre-determined. So also, the start of the forward rotation of the detaching roller is matched with feed from the feed roller. Hence, when the detaching roller starts pulling the joined fringes, a fraction of the rotation of the feed roller brings a small length of fringe in front of the nippers. The feed roller, however, stops feeding at a certain point; whereas, the detaching rollers still continue to deliver the fringe under their nip. In between, the nippers reach their forward-most position. Thereafter, they start moving back. This, together with stopping of feed from the feed roller, results in the detachment of the two portions of the fringe – one taken ahead and the other in front of the nipper bite. It may also be noted that during detachment, the top comb enters the trailing end of the fringe to comb the ending portion of the fringe. The entry of the top comb is so matched that when the joined portion of the two fringes comes under the detaching roller nip, only then, the top comb is allowed to drop into the fringe. The top comb almost remains in the fringe till the detachment is complete. Thus, the top comb also helps in the detachment of the two portions of the lap fringe.

3.1.5 WEB DELIVERY[2]

The material delivered by the front pair of the detaching roller is in the form of a thin web. Thus, with six heads, each correspondingly delivers a web in the front. The web is thin and delicate and hence, it needs consolidation before it can be taken over a long distance.

Therefore, as soon as it comes out from the front pair of detaching rollers, a small tray placed in the front helps to converge it (Figure 3.11 and 3.13). This converged mass is passed through a trumpet and is then lightly calendered by table-calendar rollers suitably placed in front of each tray and corresponding to each head. The condensed mass – comber sliver – from each head is suitably turned (Figure 3.13) at a right angle by the respective small sliver guides No.1 to No.6. The slivers thus formed are then passed over a long, polished, steel table.

The table over which the slivers slide needs to be absolutely smooth. There should be no resistance to sliver sliding over its surface. This is because; the piecing carried out in each combing cycle makes the sliver very weak.

The weakening of the sliver is also due to the high level of parallelization caused by a series of needle rows passing through the lap fringe during combing. Hence, the combed sliver on the table needs utmost care in its handling.

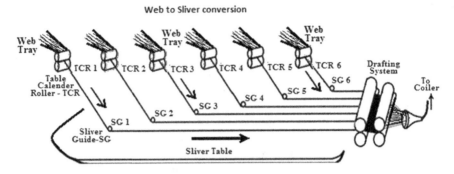

FIGURE 3.13 Web condensation, sliver doubling and drafting:[1,2]: The web delivered by the detaching rollers is condensed to form a sliver. These slivers are laid on sliver tables and subsequently drafted.

The delivery by the detaching rollers is intermittent and only a certain 'net length' is delivered in every combing cycle. Even then, the sliver journey on the table is continuous. This is achieved by suitably adjusting the speeds of table-calendar rollers to correctly match the net delivered length by detaching the roller in each cycle.

The slivers on the table finally enter a drafting system. In the old comber (Figure 3.14), the drafting system consisted of ordinary four-over-four or five-over-five drafting. The later versions, however, adopted a modified single-zone drafting system such as two-over-three.

Similarly, as against the hung weights in older drafting, there is either a spring or a pneumatic loading system. Depending upon the number of slivers, the draft is employed in the drafting system to obtain the desired weight of sliver. The sliver emerging from the drafting system (Figure 3.14) is passed through a trumpet and calendered again for more consolidation and compactness. This is very necessary as the sliver coming from the drafting system is still very weak owing to a high level of parallelization – first, due to cylinder needle action and second, due to draft in the draw box.

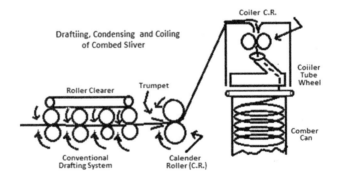

FIGURE 3.14 Drafting and coiling of sliver:[1,2] The slivers are coiled into the cans for their easy transportation to the next machine in the sequence.

It is, therefore, necessary to take maximum precaution when the sliver is finally coiled into comber cans. Further, the distance between the calendar roller and coiler calender rollers is quite large and hence, the tension draft between these two points has to be limited to not more than 1.03–105. Also, the comber cans should be equipped with springs maintained in the best condition; so that the sliver withdrawal from these cans, in the subsequent process becomes uneventful and easy.

3.1.6 Drive to the Main Parts[1,2]

The driving of the various important shafts is shown in Figure 3.15. The motor pulley drives the machine shaft through a gear. The machine shaft carries two gears – 23^T and 40^T; the former engages another gear of 90^T to drive the cylinder shaft, whereas, the latter, through the carrier wheel drives the brush shaft. Each revolution of the cylinder represents a cycle of operation which includes combing by cylinder needles and piecing-up of fringes. In each cycle, the lap fringe is gripped (or nipped) by the nippers and hence, the cylinder speed is also expressed in "nips/min". The old comber speeds were in the range of 90–100 nips/min.

With one cycle of operation, the various motions that occur during combing are required to be synchronized. For this, an index wheel, graduated with the numbers on its outer circumference, is fitted on the cylinder shaft. In old combers, the circumference of the index wheel was divided into 40 equal divisions; whereas in modern combers, it is divided into 20 divisions. The various motions are set to occur as per the index numbers. Some of them occur in chronological order; whereas some have a certain degree of overlapping.

The index wheel has a slot in which the slide is fixed with a bolt (Figure 3.16). Usually, this position is around index number 37.25. This position has to be confirmed

Driving of Important shaft

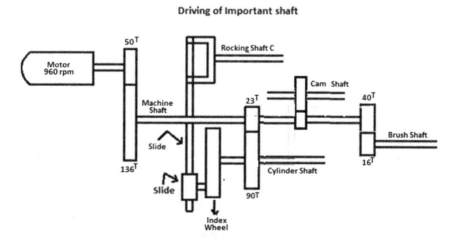

FIGURE 3.15 Drive to the main parts:[1,2] These are important shafts which are driven from the main motor. They basically control the main functioning of the combing operations.

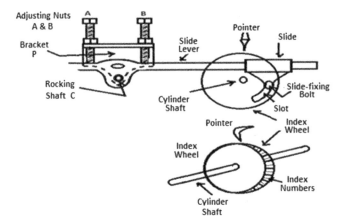

FIGURE 3.16 Motion to rocking shaft:[1,2] A very important setting in the comber. Once this setting is carried out, the timing of many other motions in combers is automatically set.

by turning the index wheel to index number 19. For this, the bolt is partially fastened and the slide is moved a little in either direction in the slot. When the slide lever starts getting its movement, the slot positions in either direction are noted and marked. The slide bolt is positioned and tightened at the centre of these two position marks.

The index wheel is turned back to index 37.25. The two adjusting bolts, A and B, with bracket P hold the position of the rocking shaft (C). Therefore, before securing the position of the slide by its fixing bolt, it is necessary to see that the two bolts – A and B – are made to touch the slide lever equally. In this position, the slide lever is expected to be perfectly horizontal. The bolts – A and B are thus secured in this position and then the slide is fixed firmly in the slot of the index wheel. The position of A and B, in this way, ensures that the rocking movement of the rocking shaft is balanced on either side during a combing cycle.

3.1.7   INDEX CYCLE[1,2]

A cycle where the various operations or events in combing take place in chronological order and at precise index numbers is called the 'index cycle'. As the index wheel is mounted on the cylinder shaft itself, it decides this order in the sequence. All the motions in the comber occur very precisely at certain index numbers. There is a pointer fixed to the framing and riding over the index wheel. For setting any event to occur at a particular index number, it is essential to first bring that number under the pointer. The appropriate setting is then carried out. The combing cycle could be imagined to be divided into two halves – one when the nippers move forward and the second when they move backward. In this respect, it becomes very necessary to position first, the nipper crank stud, which automatically decides the forward and backward rocking motion of the nippers. Further, as the main operation in a combing cycle is combing by cylinder, it would be appropriate to describe the cycle of operations starting from combing by the cylinder. Though the events described below are for conventional comber, the sequence of operations in any modern comber is almost the same. The only difference perhaps is that, whereas the index cycle in conventional is divided into 40 equal divisions, in a modern comber, it is divided into 20 equal divisions.

3.1.7.1   Events Occurring at Different Index Numbers

27 – Combing by cylinder needles starts at this index. Naturally, the fringe has to be firmly gripped at one of its ends (the back end), the other end being presented to cylinder needles. For this, both the top and bottom nippers are firmly closed to grip the lap fringe. If the grip is not very firm, the needles while carrying out combing operations would also take away the loosely held fibres along with legitimate short fibres. – an unnecessary loss of good and spinnable fibres.

During combing, the whole nipper assembly with a fully closed top and bottom nippers is made to swing back. This, to a certain extent, increases the relative speed between the lap fringe and needles. It increases the efficiency with which the needles carry out the combing operation. In addition, in old combers, the back top detaching roller is made to rock around its bottom counterpart in a backward direction. This rocking back is totally dispensed with in modern combers though.

27 to 37.25 – During this, the cylinder combing operation continues. The fringe projecting in front of the nippers is worked upon by a series of needle rows. The needles in successive rows, get finer and more closely spaced. This progressively increases the intensity of the combing action. At 37.25, the combing by cylinder needle is finished; however, the nippers are still closed. But they also reach the end of their backward swinging movement. The back top detaching roller is fully displaced backwards over its bottom counterpart (in old combers).

37.25 to 39 – At index 39, the nippers start reversing their direction of swinging and begin to move in a forward direction. The back top detaching roller still remains in its back-tilted position.

39 to 5 – The nippers continue to swing forward. Obviously, this advance is for the subsequent piecing-up operation. The bottom detaching rollers slowly turn in the backward direction to bring back a portion of the already combed fringe for piecing-up with the freshly combed one. The back top detaching roller which was, so far, in the tilted position, starts turning back to its normal vertical position around the circumference of the back bottom detaching roller. The nippers slowly start opening out and loosen their grip on the freshly combed fringe.

5 to 7 – The detaching rollers continue to rotate a little in the backward direction. The nippers are fully open and continue to advance ahead. At index 6, the feed roller starts rotating forward and feeds a small length of lap fringe for the next cycle. Owing to cylinder rotation, the needle portion on the cylinder moves around to its underside. The brush bristles placed in close proximity start working on the needles and in turn strips off the waste matter from the cylinder needles.

7 to 11 – Feeding by feed roller continues. Around index 9, the detaching rollers momentarily stop rotating back and immediately reverses the direction of rotation to deliver the material in its normal forward direction. By index 11, a few longer fibres reach the nip of detaching roller. The nippers are open and continue to move ahead.

11 to 12 – The nippers continue their forward movement. The feed roller too continues to feed the lap. The already combed fringe and freshly combed fringe are superimposed. The joined portion moves towards the detaching roller nip. The detaching of the main fringe starts. The top comb is ready behind the detaching roller line and is about to enter the fringe.

13 to 19 – The main fringe with the joined portion is brought under the nip of the detaching roller. At index 13, the top comb enters the trailing end of the fringe. The detaching rollers are in full swing to pull the fringe forward through the needles of the top comb. Some of the fibres which fail to reach the detaching roller nip after the insertion of the top comb are prevented from moving ahead. The other fibres which reach the detaching roller nip, are pulled ahead. All such fibres which are fully behind the top comb needle row or which stay partially drawn through the needles; but in any case, do not reach the detaching roller nip, remain behind at the tip of the fringe for the next combing cycle. At index 16, the feeding by feed roller stops; however, the detaching rollers still continue to pull the fringe. The nippers still continue to swing forward and move ahead towards detaching rollers. By index 19, the nippers are closest to the detaching roller and come to the end of their forward-swinging journey.

19 to 24 – The nippers start swinging backwards, at the same time, they slowly start closing. The detaching rollers still continue to pull the last remaining portion of the combed fringe. Against this, there is no advancement of the lap (feeding stopped much before, at index 16). Further, as the top comb is interposed during the detachment, only those fibres which reach detaching roller nip and which get pulled positively by the detaching rollers, are allowed to move ahead. This almost completes the 'detachment' and the portion held back by the top comb is completely detached from the fringe pulled by the detaching rollers. At index 24, the detaching rollers also stop pulling the fibre fringe and become stationary. The nippers continue to retreat backwards and are still closing.

26 – The nippers continue to move backwards and are fully closed. The new fringe gripped by the nippers is ready for the next combing cycle by the cylinder needles. The short fibres, neps, kitties, etc., which have been stopped by the top comb needle remain at the front end of the next fringe along with the waste matter belonging to the new fringe. All these matters await the cylinder needle action in the next combing cycle.

TABLE 3.1
Events Taking Place at Index Numbers in Conventional Comber

A – Rocking of Back Top Detaching Roller
12 to 27: Stationary in vertical position
27 to 37: Tilting backwards
37 to 5: Stationary in tilted position
5 to 12: Tilting back to the vertical position
B- Forward and Backward Motion of Nippers
37.25 to 19: Forward movement
19 to 37.25: Backwards
F – Top Comb operation
13 to 26: Drops down in the fringe

C – Nippers Open and Closed
7 to 19: Nippers Open
19 to 26: Closing
26 to 39: Fully Closed
39 to 7: Start Opening
D – Detaching Roller Movement
5 to 24: Forward Movement
24 to 37: Stationary
37 to 5: Backwards
E – Feed Roller Motion
6 to 16: Feeding
G – Combing by Cylinder
27 to 37: Needles Combing

27 – Again the combing operation begins. All the accumulated waste matter (short fibres neps, kitties, etc.) are picked up by the rows of cylinder needles and get consequently removed.

3.1.8 COMBING CYCLE[1]

3.1.8.1 Graphical Presentation of Combing Cycle[1,2]

A graphical representation is shown in Figure 3.18. So also, the cycle of operations is shown in two diagrams (Figure 3.17a, b) for seven different events, in all, occurring in the comber. These diagrams help in understanding the overlapping of some of the typical events occurring in a combing cycle. The order of the events taking place is usually the same.

However, in some models, there could be slight variation e.g., the detaching rollers may start rotating in the forward direction at index 6 or 7. Similarly, the top comb may be made to enter the fringe at index 12.5 or 13.5 instead of index 13. In conventional combers, the full circumference of the index wheel (360°) is divided into 40 equal divisions. whereas in some modern combers, it is divided into 20 equal parts, a matter of mere convenience.

It should, however, be remembered that, if such changes are inherent in the system, the index numbers of occurrence of all other events will correspondingly alter.

In the case of forward to backward feed or vice-versa, the index numbers of the feed roller may change. But in some combers, the timings of some events are purposely made 'early' or 'late'. These changes usually are associated with either top comb entry or starting of forward movement of the detaching roller. These are done to get certain benefits like more extraction of comber noil by more top comb influence or adjustment in the superimposition of the fringe during piecing-up operation.

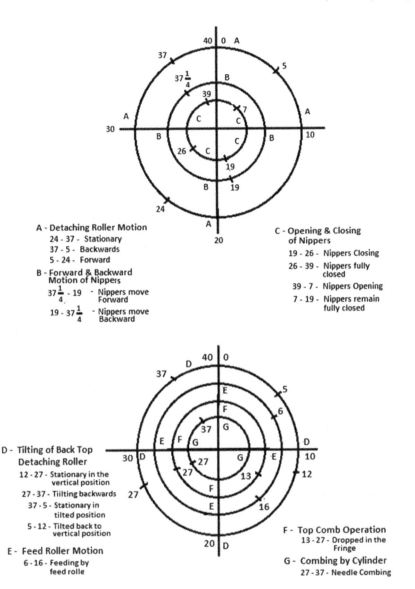

A - Detaching Roller Motion
24 - 37 - Stationary
37 - 5 - Backwards
5 - 24 - Forward

B - Forward & Backward Motion of Nippers
$37\frac{1}{4}$ - 19 - Nippers move Forward
19 - $37\frac{1}{4}$ - Nippers move Backward

C - Opening & Closing of Nippers
19 - 26 - Nippers Closing
26 - 39 - Nippers fully closed
39 - 7 - Nippers Opening
7 - 19 - Nippers remain fully closed

D - Tilting of Back Top Detaching Roller
12 - 27 - Stationary in the vertical position
27 - 37 - Tilting backwards
37 - 5 - Stationary in tilted position
5 - 12 - Tilted back to vertical position

E - Feed Roller Motion
6 - 16 - Feeding by feed rolle

F - Top Comb Operation
13 - 27 - Dropped in the Fringe

G - Combing by Cylinder
27 - 37 - Needle Combing

FIGURE 3.17 (a) Combing cycle:[1,2] All the operations in a comber are arranged in a cycle. The chart shows the timing. (b) Combing cycle:[1,2] Another set of motions such as feed roller rotation, cylinder combing operation and top comb action are also cyclic and can be represented in index cycle.

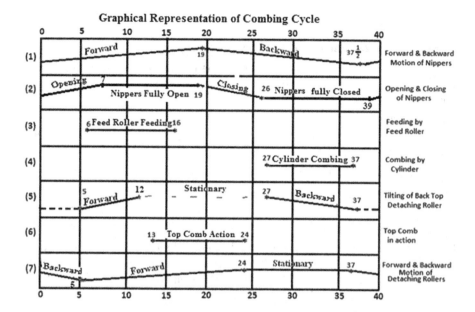

FIGURE 3.18 Combing cycle:[1,2] A graphical method is yet another way of showing the timing of different motions occurring in a cyclic manner in a comber.

TABLE 3.2
Index Numbers for Platt's and Whitin Comber Cycles

No.	Operation	Operation	Platt's	Whitin
1.	Combing Cycle	Starts	27	17.5
		Finishes	37	1.75
2.	Motion of Nippers	Starts Forward	37.5	2.75
		Most Forward	18.5	11.25
		Starts Backwards	19.5	11.75
		Backward-most	36.5	2.25
3.	Opening and Closing of	Opening – Start -	39	7.0
	Nippers	Fully Open –	07	10.75
		Closing – Start –	19.5	12.5
		Fully Closed -	26	16.5
4.	Motion of Detaching	Rotate Backwards –	39	3.25
	Rollers	End of Backward	7.5	7.5
		Motion –	7.5	7.75
		Rotates Forward –	24	13.5
		End of Forward		
		Motion		
5.	Feed Roller	Start –	6	15.25
		Finish –	16	15.75
6.	Top Comb Action	From – To	13–24	10.25–15.75

TABLE 3.3
Comparison of 2/2 and 4/4 at Varying Waste Levels[3]

Machines	19.2 % Waste Level (Mean Range %) with 4/4 2/2		15.6 % Waste Level (Mean Range %) with 4/4 2/2		7.8 % Waste Level (Mean Range %) with 4/4 2/2	
Comber Draw Box	38.2	18.3	43.5	25.2	55.5	33.6
1st DF	25.0	13.1	15.3	13.8	24.5	18.4
2nd DF	13.8	9.4	12.7	9.6	17.7	14.8

@ - New Techniques in Combing – Platt's Bul. Vol. VIII, No. 11

3.1.8.2   Index Cycle of Platt's and Whitin Comber[1,2]

In the Whitin comber, as shown in Table 3.3, and in Figure 3.19, the index cycle is divided from 0–20. Thus, the related index numbers for various events appear to be slightly different. However, the change in the index numbers is only relative.

3.1.9   MOTIONS TO VARIOUS PARTS[1,2]

3.1.9.1   Intermittent Motion to Lap Rollers

In every combing cycle, the lap rollers are required to rotate through a fixed amount. This amount is related to the feed per nip, which, in turn, is related to the effective length of the

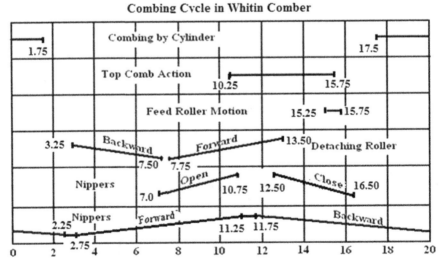

FIGURE 3.19 Combing cycle:[1,2] The first high-speed comber which came to India was the Whitin comber. In place of 40 index numbers in the conventional comber, this comber has only 20 index numbers to represent the whole cycle.

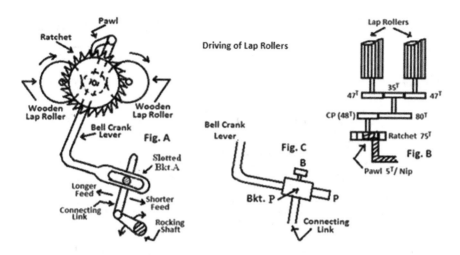

FIGURE 3.20 Intermittent motion to lap rollers:[1,2] The lap rollers turn through a very small rotation in every combing cycle to feed the lap ahead.

fibre. As the motion is intermittent, a pawl and ratchet mechanism is used. A bell crank lever carrying a pawl (Figure 3.20 A) was operated to give a certain throw to the pawl. This motion to the bell crank lever was obtained from the rocking shaft. The bell crank lever at its end carried a slotted bracket (P in Figure 3.20 C). The position of connecting link in the slot could be arranged to adjust the throw of the pawl, which in turn, varied the feed length. Alternately, in some later machines, the ratchet teeth are varied when the throw of the pawl is maintained constant. In old-conventional combers where the throw of the pawl was varied, it altered the teeth of the ratchet pushed. Thus, it was possible to push four, five or six teeth of the ratchet in every cycle (cylinder rotation).

In the Whitin comber, normally the ratchets of 16^T, 17^T and 18^T are made available for driving the feed roller. In such cases, the ratchet is pushed through only one tooth in every cycle. So also, the ratchet is directly mounted on feed roller. Thus, the ratchet with less number of teeth, results in an increase in the feed length.

The gear connections from the pawl and ratchet mechanism (Figure 3.20 B) finally lead to lap rollers, which get a small rotational motion in every cycle. In some gearings, a change pinion is provided to facilitate easy change in the feed length and to give more flexibility in varying the rotation of lap rollers. In Figure 3.20 B, it is 48^T C.P. which carries out this function.

It is very important not only to synchronize the motion of lap rollers with that of feed rollers but also to equalize it; otherwise, the lap between these two points is very likely to become slack or get stretched.

3.1.9.2 Intermittent Motion to Feed Roller

Like the lap roller, the feed roller is also intermittently driven (Figure 3.21). The ratchet is mounted on the feed roller (with a double feed roller, it is mounted on the top feed roller).

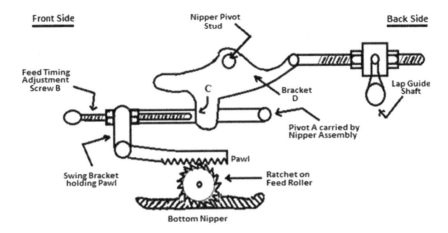

FIGURE 3.21 Drive to feed roller:[1,2] Along with the lap roller, the turning of the feed roller is perfectly synchronized. The amount of feed in both is also equal so as not to allow any stretching or slackening of the lap in between them.

The pawl is carried by the swing bracket and it is made to rest on the ratchet. The swing bracket is loosely mounted on an extension lever from pivot A which is carried by the nipper assembly. The position of the swing bracket is governed by a screw B which is used for adjusting the feed timing so that a small change can be made.

When the whole nipper assembly swings forward, both the ratchet and swing bracket also swing forward. However, the swing bracket tilts a little in the process. This brings screw B to touch the projection C of fixed bracket D. It prevents further swinging of the swing bracket. The bottom nippers carrying the ratchet, however, continue their forward movement.

The teeth of the pawl, however, still continue to engage the ratchet. As the pawl is stopped by projection C from moving further, the continued movement of the ratchet (Figure 3.22), results in the ratchet receiving a rotary motion in a clockwise

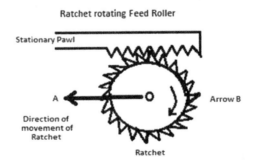

FIGURE 3.22 Turning of the ratchet:[1,2] The ratchet is mounted on the feed roller shaft. When the ratchet is moved along with the nippers, a stationary pawl resting on it helps to automatically rotate the ratchet and hence the feed roller.

direction. This happens during the forward motion of the nippers in every combing cycle. Finally, all this results in feeding a small length of lap fringe in every cycle when the nippers are moving forward. The lap fringe thus fed projects in front of the nippers. If there are two feed rollers, the bottom one is driven by the top through a gear link.

As mentioned earlier, the feed timing can be adjusted by pushing the tip of the screw either towards or away from contact point C. Though normally, the feed starts at index 6, it can be slightly varied so as to obtain early or late feeding. In yet another motion (Figure 3.23) on a different Platt's model, a very simple method is adopted to rotate the feed roller.

The ratchet carried by the feed roller is enclosed in a casing P which is centred at N. The connections from fixed links L and M are made to reach N. The pawl is connected to this linkage and is made to rest on the ratchet.

The feed roller along with the ratchet and the outer casing P is carried by the nippers. Thus, when the nippers move in the direction of arrow Q, the casing swings around the fulcrum N and this makes the pawl turn the ratchet in a clockwise direction. Both types of feeding methods discussed above take place when the nippers are moving in the forward direction. This is called 'forward feeding'.

In yet another driving arrangement, the feed roller gets its motion (Figure 3.24) again from the pawl and ratchet mechanism with some alterations. The nipper framing centred loosely on the cylinder shaft is driven forward and backward around it. The bottom nipper carrying the feed roller, the ratchet and lever N also moves forward and backward along with the nippers. During the forward movement, lever N is stopped by stop rod R which is fixed on the machine framing. As the nippers still continue to move ahead, lever N simply turns in an anti-clockwise direction (with reference to figure). This results in the pawl (carried by lever N) turning the ratchet, again in an anti-clockwise direction. Thus, the feed roller rotates and pushes the lap fringe ahead. Again, as the feeding takes place when the nippers move forward, this type of feeding is also 'forward feeding'.

On the Whitin comber, the feed roller shaft with a ratchet mounted on it is closed in a casing. The casing inside carries a pawl (Figure 3.25 A) which is made to rest

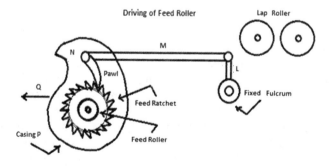

FIGURE 3.23 Drive to feed roller:[1,2] In another method, the ratchet is again mounted on the feed roller shaft. Here the oscillating motion of the pawl results in the turning of the ratchet and hence the feed roller.

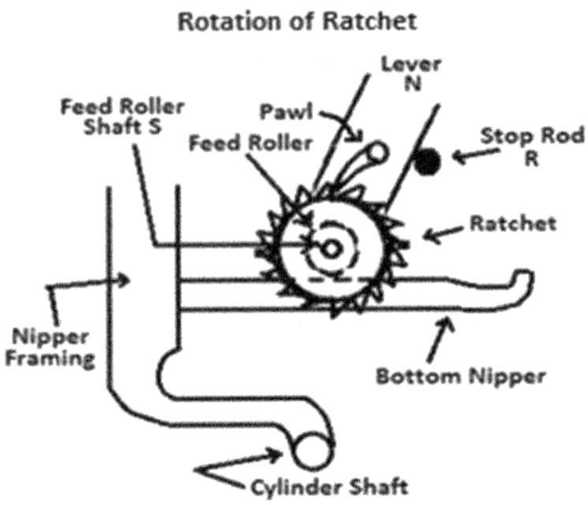

Rotation of Ratchet

FIGURE 3.24 Feed roller driving:[1,2] In this method, the ratchet is again made to rotate by a stationary pawl. At a certain point, the pawl stops, whereas the ratchet continues to move ahead. This results in the turning of the ratchet and hence pawl.

on the ratchet. Both the casing and the pawl are loosely hung around the feed roller shaft. Thus when the nipper assembly moves forward and backward, the casing and the pawl also ride along with it. Pin I is carried by the casing. During the forward motion of the nippers, the feed roller along with the casing moves forward. However, as pin I is also controlled by the curved bracket, the forward motion of the feed roller results in pin I simply turning the casing and the pawl in a anti-clockwise direction. The pawl thus rides back on the ratchet teeth.

When the nippers start reversing back (backward motion), the pin I engaged by the curved bracket, simply turns the casing and the pawl in an anti-clockwise direction, This results in the pawl turning the ratchet and hence the feed roller. As this takes place during the backward motion of the nippers, it is called 'backward feeding'.

In forward feeding (Figure 3.25 B), the casing is turned upside down. So also the position of pin I and the curved bracket is reversed. The sequence of motions takes place exactly in reverse fashion. Here pin I is turned by the curved bracket in a clockwise direction when the nippers move in the backward direction. The pawl thus rides back over the ratchet teeth.

However, when the nippers start moving in the forward direction, the curved bracket makes pin I turn the casing in an anti-clockwise direction. This results in the pawl pulling the ratchet teeth. This ultimately rotates the ratchet and hence the feed roller.

This type of feeding is called 'forward feeding'. On a Laxmi Rieter (now LMW) comber, separate ratchets are provided for forward and backward feed.

3.1.9.2.1 Pressure on Feed Rollers[2]

Pin I (Figure 3.26) can be turned in the direction of an arrow M and about fulcrum A. This pulls the spring upwards thus exerting pressure on the feed rollers.

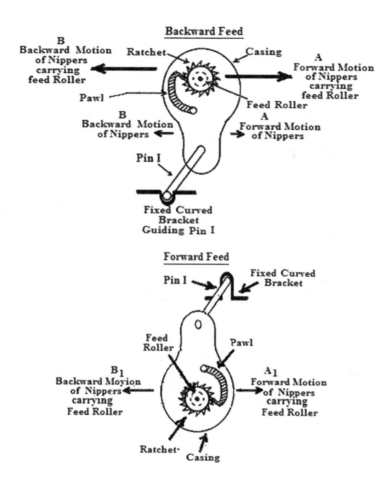

FIGURE 3.25 (A) Backward feeding:[2,5] If the feeding takes place when the nippers are moving in a backward direction, it is called backward feeding. With the same settings, when the method of feeding is simply changed, (say) from forward to backward, the waste extracted by the comber, on average, is more. (B) Forward feeding:[2,5] If the feeding takes place when the nippers are moving forward, it is called forward feeding. With the same settings, the waste extracted in forward feeding is reduced by approximately 2%.

When pin I is lowered down (present position), the pressure is released. The adjusting bracket Q is fixed by a screw N and its position along the lever P can be adjusted; e.g., if the bracket Q is taken in the direction of R (away), the point S to which the spring is attached, moves away from the fulcrum T. This increases the pressure on the feed roller.

For adjusting the position of the feed rollers with the nippers, nut-bolt C is used. When nut-bolt C is made to push lever D upwards about the fulcrum B, it shifts the fulcrum T to the left (arrow K_1). This moves both the top and bottom feed rollers nearer to the grip of the top and bottom nippers. The position of the feed roller from

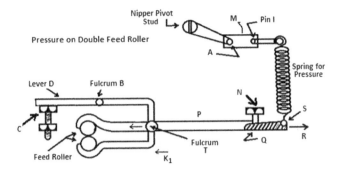

FIGURE 3.26 Pressure on feed roller:[1,2] The pressure ensures effective gripping of the lap and thus results in uniform and precise feeding of the lap during each combing cycle. It also makes certain that during detachment, when the nippers are open, the back end of the lap fringe is firmly held.

the nippers' bite is actually the feed distance and it is related to the effective length of the cotton processed. For short staples, this distance has to be closer to the nipper bite.

3.1.9.3 Motion to Nippers

The whole nipper assembly carrying the bottom nipper (or cushion plate), top nipper, top comb and feed roller is mounted on the nipper framing which is centred loosely at the nipper pivot. As shown in Figure 3.27, one of the arms of this framing is connected to one of the arms of the cradle through link A.

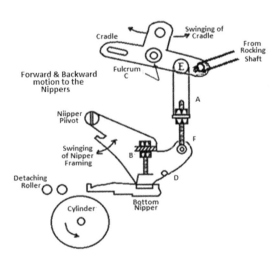

FIGURE 3.27 Motion to nipper assembly:[1,2] During each combing cycle, the nippers carrying freshly combed fringe move forward and close to the detaching rollers which bring back a small portion of the already combed fringe. The two fringes are partly superimposed.

The cradle has three arms and is centred on fulcrum C about which, it is able to swing. The bottom nipper or cushion plate is bolted to the nipper framing at D and is held securely. An adjustment is provided where the bolt screw B can lower or raise the bottom nipper in relation to the needles of the cylinder situated underneath.

Thus, whenever the bottom nipper is required to be set to the needles, both screws D situated on either side of the framing are first loosened. The setting of the bottom nipper to the cylinder needles is then carried out using screw B. It can also be easily seen that screw A (Figure 3.27), provides yet another adjustment to alter the connecting length EF between the cradle and the nipper framing.

By increasing this length, point F swings further around the nipper pivot, thus resulting in bringing the bottom nipper closer to detaching rollers.

The setting of the nipper pivot decides the coverage of the swing of the nippers. In a way, it controls the setting between the bottom nipper and back bottom detaching roller. The nipper pivot is carried by yet another bracket (Figure 3.28). As shown in the figure, one of the arms, arm A, of this bracket is centred loosely around the cylinder shaft. The other arm carries a slotted bracket which is securely bolted to the machine framing by a retaining bolt. The third arm, arm B, is connected to D through a link, C. This part of link C is threaded. The length of link C, between M and N, is adjustable and is controlled by two lock-nuts E_1 and E_2. By manipulating these two lock-nuts, the position of the nipper pivot stud can be adjusted around the cylinder shaft in the direction of arrows – F_1 and F_2. This setting is done by using a 'brass gauge' to position the nipper pivot (see Section 4.4.2). It may also be mentioned here that changing the position of the nipper pivot, changes the closing distance between the bottom nipper and the back top detaching roller when the former is closest to the detaching roller – step gauge (see Section 4.4.1).

In yet another driving arrangement (Figure 3.29), the whole nipper framing F, is loosely centred at pivot J. It is connected through a link or adjusting rod C

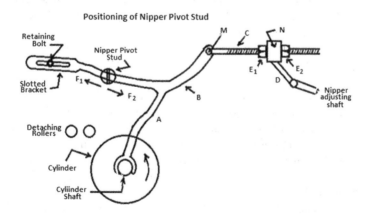

FIGURE 3.28 Nipper pivot stud:[1,2] Positioning of the nipper pivot stud ensures a fixed point around which the nippers swings to and from during a combing cycle. In a way, it also decides the most forward and most backward position of the nippers during its swinging.

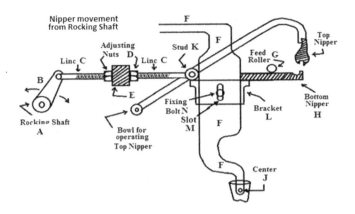

FIGURE 3.29 Motion to nipper framing:[1,2] When the framing carrying the bottom nipper is very close to its most forward position, a very important operation called 'piecing-up' takes place.

and lever B to the rocking shaft. Stud K on the nipper framing is the connecting point for link C.

As the rocking shaft moves in the direction of the arrow, link C pushes and pulls the nipper framing F around Centre J. The framing on the other side carries the bottom nipper which holds the feed roller. Thus, along with the to and fro motion of the framing, the bottom nipper along with the feed roller, also moves forward and backward. The top nipper is loosely centred on the Stud K and thus moves in a forward and backward direction along with the nipper framing.

By adjusting the length of link C, the distance of the nippers, at their most forward position can be set with respect to the detaching rollers. The bottom nipper carrying Feed Roller G is mounted on an adjustable Bracket L and with the help of Fixing Bolt N its position in Slot M can be fixed. Thus, by shifting Bracket L (up or down), the bottom nippers can be set with cylinder needles.

This helps in adjusting the penetration of the needles into the fibre fringe held by the nippers during the cylinder combing operation. In the above two arrangements (Figure 3.27 and 3.29), the motion of the nipper framing and hence that of the nippers are quite different. In the first case (Figure 3.30 A), the nippers move in the arc $A_1 - A_2$; where the distance of the nippers during their backward journey, gets progressively closer to the cylinder up to a certain point. This leads to progressive penetration of the cylinder needles. Beyond this point, however, the nippers slightly move away from the cylinder needles. This affects the combing action. In the second case, (Figure 3.30 B), the nippers move around the centre B_3 and in the arc B_1–B_2.

Their movement in this case is, more or less, concentric to cylinder needle rows. Therefore, the distance between the bottom nipper and the cylinder needles is nearly the same, during the backward swinging of nippers.

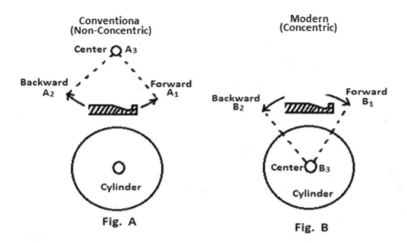

FIGURE 3.30 Movement of nippers around:[1,4,5] The old combers made the nippers move non-concentrically to the arc of the cylinder. Modern combers allow the nippers to move concentrically around the cylinder. This enables, the nippers to maintain a more or less constant distance from the cylinder

Thus, the penetration of the needles into the lap fringe presented by the nippers is almost the same throughout the combing operation.

3.1.9.4   Opening & Closing of Top Nipper[1,2]

The front arm of the cradle carries a square block called 'filbow' (Figure 3.31). A connecting rod passes through it. The rod carries latch M at its lower end. This latch, when

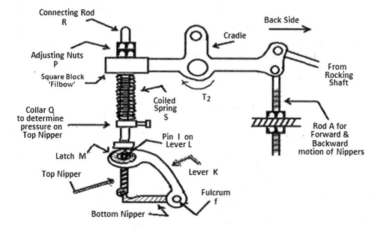

FIGURE 3.31 Opening and closing of top nipper:[1,2] The movement of the cradle received from the rocking shaft allows the nippers to open and close during their forward and backward motion respectively. The nippers are closed during combing, whereas they are open during detachment of the lap.

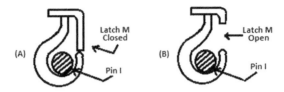

FIGURE 3.32 Locking of top nippers:[1,2] The top nipper in the normal course is controlled by a filbow lever. Opening and closing of the nipper is effected only when the latch is closed around the pin. A: Latch Closed; B: Latch Open.

locked, holds pin I (also shown in Figure 3.32 A and B) carried by lever K. The lever K is fulcrums at 'fulcrum f' and holds the top nipper. While locking latch M with pin I, the former is lifted up and then inserted around the pin. After inserting the pin, the latch is lowered again, thus locking the pin (Figure 3.32 A and B). When the pin is locked in this way, the top nipper becomes one piece with the vertically extended connecting rod R.

The rod R also carries a coiled spring S guided on one side by the filbow and on the other side by a collar Q. The pressure on the spring can be adjusted by positioning the collar on the rod. For example, if the collar is taken-up (with reference to Figure 3.31), the spring gets compressed, thus exerting more pressure on the collar, which in turn applies this pressure on the top nipper.

In addition, when the nipper assembly moves back during combing by cylinder, this spring is further compressed, thus exerting higher pressure on the top nipper during cylinder combing. Owing to this, the lap is firmly gripped during the combing by the cylinder. Switching over from lighter laps to heavier laps necessitates an increase in this pressure. For this, in some cases, the pressure can be increased by simply pulling collar Q upwards. In some typical cases, a more powerful spring needs to be used.

It is evident (Figure 3.31) that when the nippers move forward, the movement of the cradle in the direction of A_2 raises the front arm of the cradle. Ultimately, this raises the connecting rod R and hence the top nipper around fulcrum f. The schematic diagram (Figure 3.33) reveals another interesting thing. With forward motion,

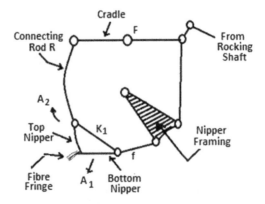

FIGURE 3.33 Schematic diagram of nipper framing:[2,5] Such a diagram helps in understanding the lever connections and their movements in opening and closing of the top nipper.

the nippers move around the nipper pivot and hence their motion is in the direction of arrow A_1; whereas the top nipper movement, effected by the cradle takes place around point 'f'.

This results in the lifting of the top nipper from the bottom nipper. Thus, the link connection of the top nipper to point 'f' through link K_1 is very important. It gives the top nipper an extra margin of lifting during the forward motion of the nippers. When the nippers move forward, the opening that the top nipper provides, should not be less or more than a certain precise distance. With the lesser opening, there may be an obstruction for the free movement of lap fringe during detachment; whereas with more distance, there would be unnecessary lifting-up of the top putting more strain on the mechanism, especially the top nipper spring.

Also, the extra lifting would compel the top nippers to travel back this extra distance before closing for cylinder combing operation. It may also lead to its jerky closing. This gap when the nippers are fully open can be adjusted by adjusting nut P, (Figure 3.31) on connecting rod R.

In yet another arrangement (Figure 3.34), the top nipper is centred at N. The rod R carries at one of its ends, the top nipper whereas, at its other end, is bowl B controlled by piece K. The pin I at the centre of the bowl is connected through a powerful spring S to the machine framing. The spring S holds bowl N in firm contact with the lower profile of piece K.

Owing to the typical shape of this profile the bowl is pressed downwards when the nippers move forward. This lifts the top nipper placed at the end of the rod R. The extent, to which this raising is required, can again be controlled by adjusting the position of piece K which can be lowered to raise the top nipper further.

During the backward motion of the nippers, however, bowl B passes back and past the point 'T' on piece K.

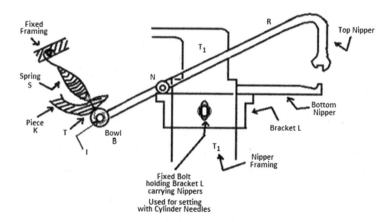

FIGURE 3.34 Pressure on top nipper:[1,2] The bowl mounted at the back end of the lever holding the nipper is pushed down, thus opening the top nipper. The use of a powerful spring helps in giving adequate pressure on the top nipper during combing by the cylinder.

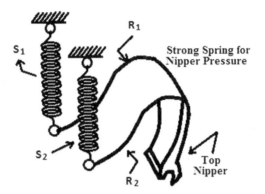

FIGURE 3.35 Pressure on top nippers:[1,2] Very powerful springs are attached to the trailing ends of the top nippers. During the combing operation, the springs are in the stretched condition, thus giving adequate pressure on the nippers.

There onwards, the spring pressure starts acting on bowl B and pulls the bowl upwards. This closes the top nipper. In their closed position, the full pressure of spring S is exerted on the top nipper, thus enabling them to hold the fibre fringe very firmly.

There are two options for varying the pressure on the top nipper. Either a new powerful spring can be used in place of an older one or the spring can be stretched further to increase the pressure on top nippers when they are closed. A small variation in the spring pressure is thus possible (Figure 3.35). When a top nipper opens in every combing cycle, the spring gets regularly stretched. Over a number of years, therefore, it is likely that the spring becomes slack and weak. At this point, it is always advisable to put on a new spring. In Figure 3.34, only one spring is shown. Actually, there are two such springs – S_1 and S_2 – on either side of the top nipper extensions R_1 and R_2 (Figure 3.35).

It is very important to equalize the pressure of these two springs. It helps in ensuring uniform nipper bite across the width of the nippers. If the spring pressure is unequal, the nipper bite also becomes uneven across the nipper's width. This leads to good fibres being lost due to inadequate nipper grip. This can be easily noticed by observing the comber noil deposited on the aspirator perforated area.

3.1.9.5   Motion to Detaching Rollers[1]

As stated earlier, the detaching rollers have three states of motion. They rotate backwards to partially bring back the already combed lap fringe for piecing-up with newly combed fringe. They then rotate forward to deliver the pieced-up fringe ahead

During combing, however, they remain absolutely stationary. All these motions are initiated from Sector S (Figure 3.36 A), which is loosely fulcrumed at fulcrum F. The sector is guided by Cam M on the Cam Shaft B which gets its motion from the machine shaft (Figure 3.36 B). With the motion of Cam M, the sector is moved in the direction of arrows T1 and T2 (Figure 3.36 A and C). The rocking of the sector about fulcrum F, rotates the wheel W in either direction. Thus, if the sector rotates in

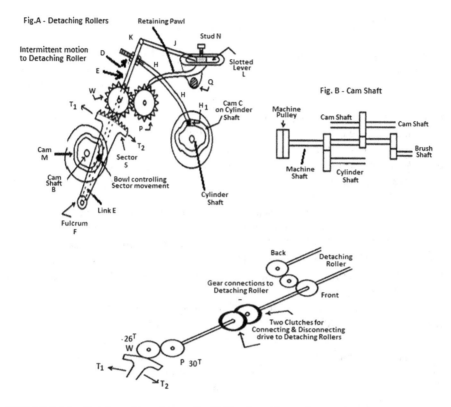

FIGURE 3.36 Driving of detaching roller:[1,2] The detaching rollers are intermittently rotated backward and forward to piece up the joint of old and freshly combed fringe and deliver the combed sliver ahead.

the direction T1, the wheel W would rotate in the clockwise direction and vice-versa. The motion of the wheel W in either direction is finally conveyed through wheel P (Figure **3.36 C**) to the detaching rollers. Thus, when the sector rocks in one direction, the detaching rollers rotate in the backward direction. When the sector reverses its rocking, the detaching rollers start moving in the forward direction.

It may appear that, as the rocking of the sector is equal in either direction, the detaching rollers would equally rotate in a backward and forward direction. However, though this rocking of the sector is equal, the motion finally passed on to detaching roller is never the same. It has been already mentioned that the difference between the forward and backward rotation of detaching roller results in 'net length' delivered in the forward direction.

Thus, [Forward Rotation – Backward Rotation = Net Equivalent Length Delivered]

The reduction in the backward rotation of the detaching rollers is brought about by yet another cam, cam C (Figure 3.36 A), mounted on the cylinder shaft. This cam controls the bowl H_1, which is connected through H, to a link E with the help of adjusting bolt D. The meshing of W and P can be adjusted with the help of this adjusting bolt D.

So as to rotate detaching rollers in the backward direction, the sector swings in one direction. Accordingly, the wheel W is also rotated. However, all its rotational motion is not transferred to wheel P. This is done by simply disengaging the connection between W and P, which cam C does. This cam keeps the wheel P disengaged for the first part of the sector movement. After this, the wheel P is made to engage wheel W. The remaining movement of the sector is thus transferred to detaching rollers, which rotate only partly in the backward direction. When the sector completes its throw, it reverses its swinging. The wheel P continues to be engaged. Reversing of the sector alters the rotational direction of wheel P and hence that of detaching rollers so as to rotate them in a forward direction. At the end of the forward rotation of the detaching roller, the sector also reaches its other extreme position. At this time, again the wheel P is disengaged. This is around index 26. It may be noted that immediately after this index 26, the combing by cylinder starts. Hence, the detaching rollers are required to be stationary. The disengagement of the wheel P with wheel W brings about this state of detaching rollers.

The wheel W is carried by the link E which is centred on the fulcrum F at the bottom (Figure 3.36 A). This link is supported by a connecting rod J at the top. The rod carries stud N. The stud can be adjusted in the slotted lever L, which is loosely centred on Q. The adjustment of stud N in the slot brings about the correct positioning of pawl K on the wheel P. This pawl is a retaining pawl and it holds the wheel P securely when wheel P gets disengaged from wheel W.

As mentioned, when cylinder combing takes place, the detaching rollers must remain stationary. The pawl holds the wheel P firmly and steadily during this time and does not allow the detaching rollers to have any motion.

Another simple driving of detaching rollers is shown in Figure 3.37 A. The sector S is mounted loosely at 'fulcrum F' and carries a bowl B. The groove in the cam placed on the cylinder shaft regulates the motion given to the bowl B, and therefore, the movement of the sector S. When the cam rotates, the sector is made to swing in either direction – T_1 and T_2 (shown by arrows). When the sector swings, the sector teeth engaging another pinion U rotate it in either direction and this is transferred to pinion N. In turn, the motion is ultimately passed on to detaching rollers through clutch W (Figure 3.37 A) During the forward rotation of the detaching rollers, the clutch is made to fully engage (Figure 3.37 B. When the pinion U receives full rotations through the sector, it transfers them to the detaching rollers. However, during combing by the cylinder, there is declutching and no motion is given to detaching rollers. During this time, however, the sector still continues its reverse swing.

Had the grooved cam been designed to give full throws to the sector in one direction (forward rotation of detaching rollers); and immediately thereafter, the groove in the cam designed to impart no motion to the sector (cylinder combing operation), even then, bringing the sector back to its original position for the next cycle would become a serious problem.

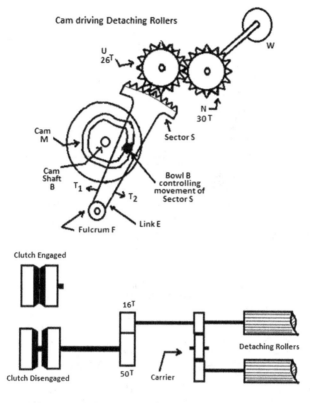

FIGURE 3.37 A: Drive to detaching rollers:[1,2] Forward and backward motion to the detaching roller is imparted through the bowl and the grooved cam. B:[1,2]: Engaging and disengaging of clutch.

Around the late 1960s, the Platt brothers introduced their high-speed comber. They had a different mechanism (Figure 3.38 A) to rotate the detaching rollers. There are two compound gears – internal (A_2) and external (A_1). The former is loose on the cylinder shaft. The latter is also loose and connected to the detaching rollers through 56[T] wheel.

The internal gear partly engages with 80[T] gear loosely riding on an eccentric E whereas, the eccentric is fast on the cylinder shaft. This 80[T] gear is connected through – X_1, X_2 and bell crank type connection 'f' to link connection P (Figure 3.38 B). The lever L is held loosely at its bottom end and is also connected to a grooved cam B by another link R.

This grooved cam is fixed on the cylinder shaft. This cam initiates an up and down movement of lever L (Figure 3.38 C). This operates X_1 and X_2 and causes the 80[T] gear to move in the direction either – clockwise or anti-clockwise. The swinging rotation of the 80[T] wheel is obviously equal in both directions. However, the contact between the 80[T] wheel and internal gear is controlled by the eccentric E (Figure 3.38 C).

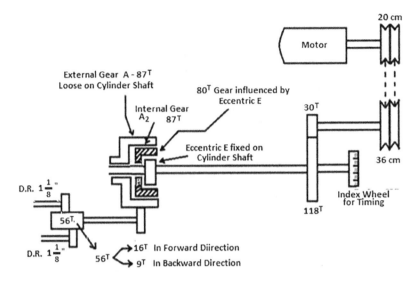

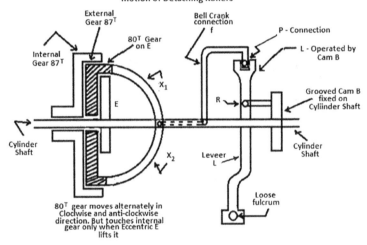

FIGURE 3.38 A: Forward and backward motion of detaching rollers (century comber):[1,2] Platt introduced their then high-speed comber running around 150 nips per minute. The driving of detaching roller was done ingeniously by intermittently disconnecting the gears instead of rocking the sector and cam. B: Importance of internal and external gears:[2,6] The drive through internal gear is effected by alternating the motion to one of the wheels. Here too, the contact is partially disconnected in one of the directions. C: Driving of 80T gear:[2,6] This wheel is made to oscillate equally on either side. However, in one of the directions, only a part of the motion is conveyed to detaching rollers.

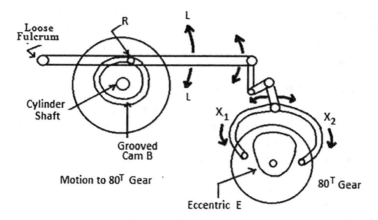

FIGURE 3.38 (CONTINUED)

When the 80^T wheel moves in one direction, the eccentric establishes its contact with the internal gear and rotates it. The external gear being compounded with the internal gear, thus also rotates. This external gear, as mentioned earlier, is connected with the detaching rollers, which are then rotated in one direction (forward delivery) during full swinging rotation of the 80T wheel. Immediately, thereafter, the eccentric E breaks the contact of 80T with the internal gear.

Though the swinging movement of the 80^T wheel is continued in the reverse direction owing to cam B, the broken contact between the 80^T wheel and internal gear, does not pass the motion, through the internal gear to the external gear and hence the detaching rollers remain stationary during cylinder combing operation.

During the later part of swinging of 80^T in the reverse direction, eccentric E again establishes the contact between the 80^T gear and the internal gear. The external gear, compounded with the internal gear thus conveys this motion to the detaching rollers, but in the reverse direction for the remaining part of the reverse swinging rotation of the 80^T wheel.

In the next cycle, again the swinging of the 80T wheel starts in the normal direction. Here again, the contact between the 80^T gear and internal gear is maintained. The detaching rollers, after rotating only partly backwards, thus start moving in the forward direction to feed the combed material in the web form.

Thus, an eccentric E is mainly responsible to break and establish the contact between the 80^T gear and the internal gear. When the contact is established the motion is passed on to detaching rollers through external gear. Whether the detaching rollers rotate in the forward direction or backward direction depends upon if the

80^T gear is rotated by fork levers – X_1 or X_2 – in the normal direction or in the reverse direction. It may also be noted that when an eccentric breaks the contact between the 80^T gear and internal gear, no motion is transferred to detaching rollers i.e., they remain stationary. This happens during combing by cylinder.

3.1.9.6   Rocking of Back Top Detaching Roller[1,2]

In a conventional comber (old Nasmith comber), the back top detaching roller was made to rock back around its bottom counterpart. This rocking motion was given for a specific purpose. The movement was timed during combing by the cylinder. The basic idea of this tilting of the back top detaching roller was to protect the tail end of the already combed fibre fringe from whirling air currents generated around the cylinder by the cylinder needles during combing. In a way, it also helped better superimposition

The rocking shaft S initiates this swinging action. As shown in Figure 3.39 A, the rocking shaft moves in the direction N_1 and N_2. The rocking shaft itself derives its motion from yet another cam on the cylinder shaft (Figure 3.39 B). The rocking motion of the rocking shaft in the above directions is conveyed to point P on the forked lever J, through a connecting rod R (Figure 3.39 A and C). The forked lever J is fulcrum at 'f', which also carries a split clamp.

The rocking shaft S gets its rocking motion from the cam on the cylinder shaft. This motion through clamps is transmitted to the sleeve of the back top detaching rolled via connecting link M (Figure 3.39 A). This is responsible for the swing back of the top detaching roller. Though in the figure only one link (M) is shown, there are two such links holding the sleeves of the back top detaching roller on either side.

A powerful spring is connected to an extension M' of link M. The other end of the spring is connected to a metal bracket K at K'. The bracket carries a slot at the top to hold the top clearers for detaching rollers. The spring provides adequate pressure on the back top detaching roller and helps in improving its grip on the fleece during detachment. The two nuts P_1 and P_2 provide an arrangement for variation in the pressure. Similarly, the two nuts T_1 and T_2 provide a setting arrangement for adjusting the position of the back top Detaching roller for its backward tilting.

It is very important to see that the back top detaching roller, in its extreme back tilting position, does not touch the top comb needles; otherwise, its surface is likely to get damaged.

In old combers, this rocking commenced when combing by cylinder started at index 27 and finished at index 37. This helped in keeping the last portion of the already combed fringe under its nip so that the back end of the fringe was not disturbed.

During index 37 and 5, the back top detaching roller remains stationary in its tilted position. Thereafter, it again starts rocking forward to its normal, vertical position. This tilting did seem to be much more useful in helping the piecing-up operation by rolling down the already combed fringe onto the top surface of the back, bottom detaching roller. It seemed to help in a proper superimposition of freshly combed fringe riding over the top of already combed fringe (Figure 3.40 – fringe B riding over fringe A).

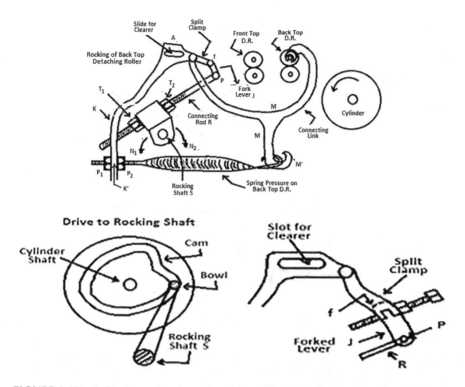

FIGURE 3.39 A: Rocking of back top detaching roller (D.R.):[1,2] All old combers had their back top detaching rollers swinging around their corresponding bottom detaching roller. B and C: Back top detaching roller and clearer connection:[1,2] The rocking of the rocking shaft results in various states of motion to the back top detaching roller. The rocking also initiates motion to the clearers which clean the detaching rollers.

The rocking back of the top detaching roller also helped the joint of old and new fringes to come under the detaching roller grip a little earlier. When the back top detaching roller rocked forward to restore its normal position, the top comb entered the fringe. This further helped in the closer entry of the top comb with reference to the nip of back detaching rollers. In addition to this, the nip movement in the forward direction caused by the forward rocking of the back top detaching roller also regulated the drafting during detachment. In modern combers, the projecting tail of already combed fringe at the back of detaching rollers is very short and hence, this rocking mechanism is totally dispensed with.

3.1.9.7 Top Comb Action[1,2]

The top comb derives its motion from the nipper assembly. The top arm of the cradle is connected to bracket V at V' (Figure 3.41A and 3.41B). With the forward motion of the nippers point V' is pulled back (arrow **1**) about fulcrum F, thus shifting the

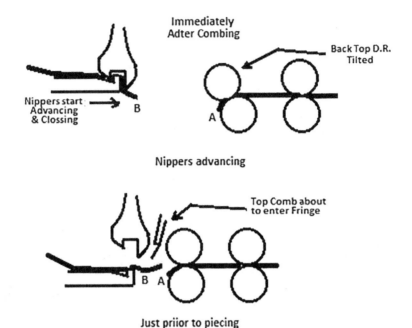

FIGURE 3.40 Rolling down of earlier combed Fringe:[1,2] The laying down of the earlier combed fringe on the surface of the bottom detaching roller enables perfect superimposition of the two fringes.

slotted end of V in the forward direction (arrow **2** – Figure 3.41 A and 3.42. The top comb has a projection block S' (Figure 3.41 A) and it is guided in the slot S of bracket V (fig. 3.42), when it is mounted.

The top comb has a bowl B (Figure 3.41 A) fixed in the slot 't' and is adjustable. This bowl, during cylinder combing operation, is made to rest on the platform D.

The platform D is fixed on the machine framing. Yet another bowl, P, on bracket V (Figure 3.43) is a stationary bowl and merely guides the slot P' of the top comb during its raising and lowering motion.

3.1.9.7.1 Motion to Top Comb[1,2]:

The top comb moves into the fringe during the piecing-up operation and is withdrawn before the start of the cylinder combing action. Both these movements are carried out by neither any cam nor by any lever connections.

When the nippers go back for cylinder combing, the cradle swings in an anticlockwise direction (Figure 3.42), thus pushing the point V' of bracket V around its fulcrum in an anti-clockwise direction. The top comb which is carried by this bracket also swings in the same direction. But just prior to combing by cylinder, the bowl B on the top comb touches the stationary platform D (Figure 3.41 A and 3.43) and this does not allow the top comb to swing further. The bracket V, however, continues to swing anti-clockwise. Therefore, this movement of bracket V, in trying to

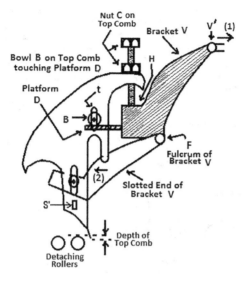

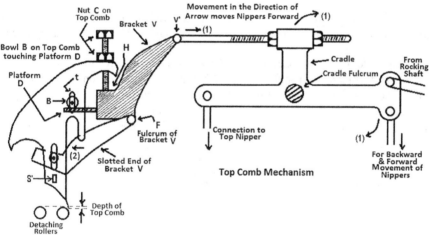

FIGURE 3.41 A: Top comb:[1,2] The main object of the top comb is to comb the training portion of the fringe being detached. Though it remains passive during this action, it is responsible to restrict the flow of unwanted fibres from going into slivers. B: Connections to top comb:[1,2] The entry of the top comb into the fringe must be withheld till the joint portion of the two fringes comes under detaching roller nip.

swing further, results in raising the top comb upwards (against platform D) and out of the line of lap fringe. During the forward motion of the nippers (cradle swinging clockwise), till index 12, the platform D maintains its contact with bowl B of the top comb and hence, the top comb still remains in a raised position.

But immediately after index 12, the bowl B on the top comb assumes a position in which it is just about to leave its contact with the platform D. The bracket V also moves

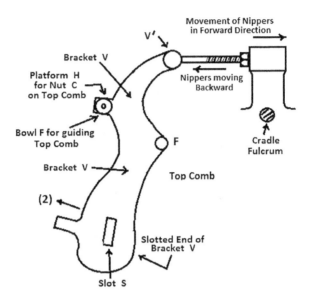

FIGURE 3.42 Top comb adjustments:[1, 2] The top comb is required to be adjusted for its effective functioning, In a traditional comber, the top comb movement was controlled separately.

around the fulcrum F (Figure 3.41 A and 3.42) in a clockwise direction and lifts the top comb bowl further away from Platform D. This allows the projection block S' (Figure 3.41) of the top comb (being guided by slot S of the bracket V) to slide down.

The top comb thus drops down into the lap fringe. This makes the top comb needle row come only a little below the line of the lap fringe (Figure 3.41) which is

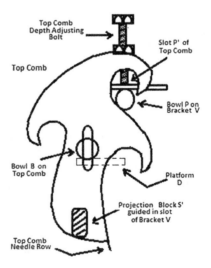

FIGURE 3.43 Top comb accessories:[1,2] These are required for the top comb to perform at a precise time with respect to the combing cycle.

being detached. This is called 'top comb in action. It must be mentioned here that the top comb action, at any cost, must not start before the joint of the pieced-up fringes comes well under the detaching roller grip; otherwise, the top comb would partly hold back some of the good fibres from the freshly combed fringe, thus seriously leading to loss of good fibres.

In yet another mechanism (Figure 3.44), the motion to the top comb is obtained in a very simple way. The top comb F is carried by a bracket T which is bolted to lever G by a fixing nut, M. The lever G is centred at Q, on an extension of nipper frame H.

Therefore, when the nipper frame moves forward and backward, the lever G, carrying the top comb, also moves along with it. On the lever G, there is a bowl N fixed in the slot. The bowl can be adjusted in the slot. The bowl N, rests on the plane of an inclined arm P. The lever G is also provided with adjusting screw K which rests on the extension of nipper frame H at H'.

During the forward motion of nippers, the lever G and the bowl N slide over the inclined plane P. Around index 12, the bowl N is at the tip (as shown in Figure 3.44) of the inclined arm P. Any further movement of nipper frame H and hence that of lever G, results in the bowl losing contact with the inclined arm P. In fact, it starts slowly moving down. This allows the top comb to slowly drop down into the fringe.

However, the screw K, at some point of the time later, touches the extension H' and regulates and restricts the top comb fall. In fact, the setting of K decides how much the top comb should fall down into the lap fringe. The bracket T, holding the top comb, further provides the necessary adjustment for (a) moving the top comb

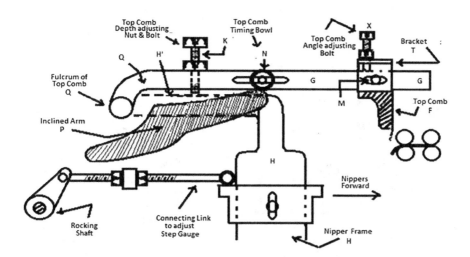

FIGURE 3.44 Motion to top comb:[1,2] Various arrangements of setting the top comb for its effective action in combing the trailing end of the fringe are provided. Thus, the top comb during detachment of the fringe is simply made to enter the ongoing joined fringe.

towards the detaching rollers (bracket M can be moved in the slot) and (b) correcting the inclination of the top comb (bolt X can be adjusted). During the return journey of the nipper frame (backwards motion), the bowl N is again lifted-up by an inclined arm P and thus the top comb is out of action. This happens a little prior to combing by cylinder needles.

In modern combers, the elaborate arrangement discussed above is totally dispensed with. The top comb in the Whitin forms the part of nipper framing and moves along with it (Figure 3.45). A bracket C holding the top comb is slotted and bolted to the nipper framing by bolt B. This slot carries a recess in which the top comb is held securely. By loosening bolt B, the bracket C, and therefore the top comb, can be adjusted with respect to detaching rollers. Further, by loosening the check nut D, an adjustment can be made by screw E to lower or raise the top comb.

This adjusts its depth of penetration into the lap fringe during detachment. Due care, however, has to be taken while lowering the top comb. This is because; the top comb fits in the recess a little snugly and its lowering is effected by raising the screw E. This may not allow the top comb to freely slide down. While lowering the top comb therefore, a little tapping from the top is required to ensure that it is lowered down according to the movement of screw E. Thus, the projection of E, in this new case, must touch the bracket C so as to ensure that the top comb does not slide further below, during working.

In Platt's century comber (Figure 3.46), the top comb is not carried by the moving nippers frame. In Figure 3.46, a separate bar 'f' receives partial rotational motion from an eccentric through link L. This gives a rocking motion to the link and hence to the bar 'f' in the direction (R_1–R_2), R_3 and (R_4–R_5) in succession.

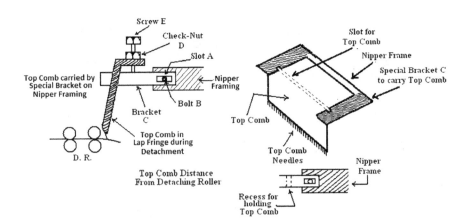

FIGURE 3.45 Modern top comb:[2,7] The top comb on a modern comber becomes a part of the nipper framing and moves in an arc, forward and backward along with it. There is no separate motion needed to make the top comb operative or inoperative.

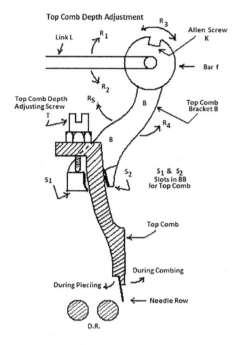

FIGURE 3.46 Top comb Platt's century comber:[2,6] The motion to the top comb is given in a very ingenious way. The oscillatory motion given to the top comb brings it into action. When it swings away, it is automatically out of action.

This gives motion to the top comb, where, it is taken nearer to and away from detaching roller.

However, owing to the angular disposition of the comb, when it moves away (R_4), it is out of action. Similarly, when it moves nearer (R_5), it is automatically brought closer to detaching roller for carrying out its normal operation of combing the tailing end of the fringe. The depth of the top comb is adjusted by screw T, whereas the distance of its needles from the detaching rollers can be adjusted by positioning the bar f with the help of Allen screw K.

3.1.10 BRUSH SHAFT AND ASPIRATOR[1,2]

During cylinder combing action, the fine needle rows extract waste, as they pass through the lap fringe. This waste is comprised of short and immature fibres, kitties and neps. After each combing operation, these needles are thus required to be cleaned during every combing cycle.

The brush shaft is positioned just below the cylinders. Accordingly, the brushes are set so that their bristles just penetrate a little into half-lap needling (Figure 3.47). The brush shaft gets its drive from the machine shaft. The brushes come in two halves and are assembled on this brush shaft at appropriate positions.

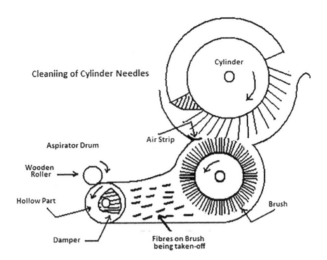

Cleaniing of Cylinder Needles

Cylinder

Aspirator Drum

Air Strip

Wooden Roller

Hollow Part

Brush

Damper

Fibres on Brush being taken-off

FIGURE 3.47 Brush and aspirator:[1,2] In every cycle, the cylinder needles remove short fibres and some other foreign matter. The brush removes it from the needles and ultimately gets cleaned by a powerful suction from the aspirator

The setting point is provided for adjusting the penetration of brush bristles into cylinder needles. The adequate penetration of the brush bristles into cylinder needles ensures a complete stripping and cleaning of the needles. However, inadequate penetration or too much penetration are both undesirable. With the former, the needle cleaning process is unsatisfactory; whereas with the latter, there is excessive wear of the bristles. or even damage to the needles in the case of half-lap needling.

Incidentally, in spite of the careful setting of the bristles, the wearing of bristles is still unavoidable owing to friction between the bristles and the cylinder needles. After a certain period, therefore, the bristles do wear off and subsequently fail to carry out their job. After regular observations on both the noil extracted and the cleanliness of cylinder needles, it is possible to deduce whether the bristle penetration is adequate. When bristles, owing to wear, fail to penetrate the cylinder needles, it becomes necessary to raise the brush shaft closer to the cylinder needles. However, there is a limit up to which the raising of the brush shaft can compensate for the reduction in the diameter of the brush bristle. This is because the bristles start losing their flexibility and steadily become a little rigid below a certain height. This is more troublesome, as they are likely to start putting a little extra pressure on the needles in the case of old half-lap segments during cleaning. Subsequently, it may lead to needle breakages. At this point in time, it becomes advisable to replace the old worn-out brushes by using a fresh brush set.

This may also become essential when the observation reveals that there is uneven wearing of brush bristles. If there are two or three such brushes where this extra wearing of bristles is observed, replacing only those brushes is also not the correct solution. This is because, when the new brushes are put in place of only those which are worn-out, still it will lead to an uneven diameter of brushes over the bristles. In this state, therefore, it is not possible to set the brush shaft correctly. Using a totally new set of brushes for all the heads thus becomes the only logical solution.

The brush bristles, in turn, require a subsequent cleaning and this is carried out by an aspirator (Figure 3.47), which is placed at the back side of the comber machine. It is a long, extended, hollow, perforated drum which is designed to extend and cover all the heads The drum contains a damper – a hollow pipe (Figure 3.47) – situated inside it. This hollow pipe is fixed inside the drum. The hollow pipe has square holes cut on its surface. The cut portions are made to face the corresponding brushes of each of the heads. Whereas the outside perforated drum is rotated intermittently in each cycle, the inner hollow pipe (damper) is kept stationary in its position (Figure 3.48).

This hollow pipe, at the centre, is linked to a duct which, in turn, is finally connected to a suction fan. The air is drawn by the fan through the duct, the hollow pipe and finally through the fine perforations of the damper. The suction is finally made to reach the brush shaft zone. This creates an exhaust around the brush region. The exhaust demands an air supply which is provided by the surrounding air. A small air

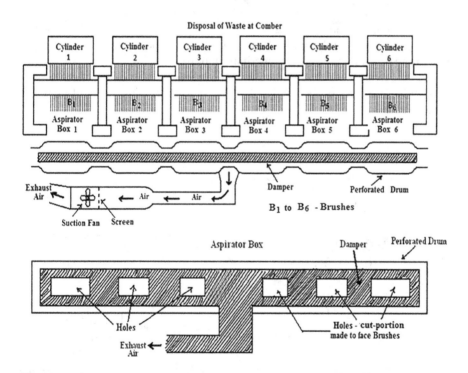

FIGURE 3.48 Suction through aspirator:[1,2] The aspirator with a circular, perforated hollow drum is connected to a powerful suction fan. The air coming from the brush is sucked in. The perforations prevent the fibres from entering the drum. The deposited fibrous matter over the drum surface is called comber noil.

strip is provided very close to the brush bristles and provides the necessary impetus for the strong air currents to move around the brush and strip the brush bristles. This helps in cleaning the fibrous and other matters, picked up by the bristles from cylinder needles.

During every cycle, the brushes clean the cylinder needles when the respective half laps face the brush bristles. For more powerful and effective cleaning, the speed of the brush shaft is almost more than 2.5 times the speed of the cylinder shaft.

The deposited thin fleece of waste over the tiny perforations of the aspirator is subsequently taken around in a sheet form over and around the aspirator drum and is deposited in the waste containers which are appropriately positioned at the back and under the aspirator.

3.1.11 WEB CONDENSATION AND DRAWING[2]

The web delivered by the detaching rollers is a semi-transparent film of fibres (Figure 3.49). Further, as mentioned earlier, it has periodic joints that are spaced at regular intervals (piecing operation). Therefore, even when the web appears to be continuous, it is not so. Depending upon the superimposition*, these joints can be seen with close observation.

The electronic evenness testing machines can spot these joints and show them as waves appearing at regular intervals. It is either a prominent thin place (less superimposition) or a thick place (over superimposition).

The best joint, however, shows minimum deviations from either of these situations. The wavelength of such periodic variation, as given in the form of graphs by these instruments,

Web condensation and Calendering

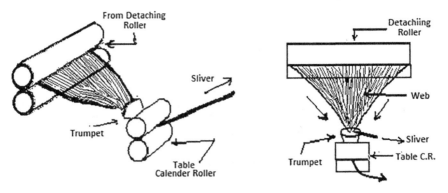

FIGURE 3.49 Condensation:[1,2] The web delivered by the detaching rollers is very thin, almost semi-transparent and weak. Especially the intermittent piercing joints make it really weak. It requires immediate condensation.

---

* Overlapping of two fringes – already combed and freshly combed fringes.

clearly shows peaks at intervals with a wavelength equal to 'the net delivered length by the detaching rollers in every cycle. The web delivered is thus very weak.

Hence, it requires immediate condensation, obviously into a sliver form. A trumpet placed in front of detaching roller does this job. There is some distance between the detaching roller and this trumpet and during this journey, the web is guided by a supporting tray underneath.

The draft level in this region needs to be an absolute bare minimum; so that there is no undue stretching of the web. Any excess in this level adds to the irregularity in the comber sliver.

The extent to which the two fringes are superimposed can be controlled (Figure 3.50 – A and B). This is done by adjusting the timing of forward rotation of detaching roller. Even with such adjustments in the timing, the joint is never perfect. This is because the fibre in the two fringes – one already combed and the other freshly combed – trail-off randomly. Hence, the perfect matching of the ends of the two fringes is not possible. Further, the adjustment in the timing of detaching roller cam has to be done manually and again there is element of some limitation. This makes the occurrence of piecing wave unavoidable, though its impact (amplitude of wave) can be minimized by appropriate forward timing of detaching rollers. The evenness tester shows the irregularity due to piecing wave. With the best timing of starting of forward rotation of the detaching roller, the height of the irregularity peak at the .joint is remarkably reduced.

3.1.11.1 Symmetric and Asymmetric Condensation

It would be very interesting to note that the method of web condensation plays an important role in deciding the height of the irregularity peak. The conventional combers had 'symmetric web condensation' (Figure 3.51 A), where the piecing wave was very prominent. In this type of web condensation, the joints run in a perpendicular plane to the direction of the sliver formed. This localizes the joints across the sliver and at regular

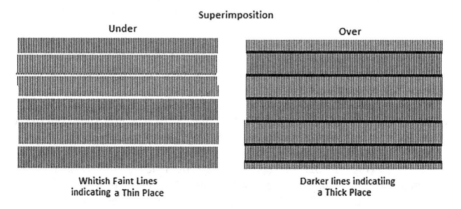

FIGURE 3.50 A: Under and B: Over superimposition:[2,5] The over lapping of already combed fringe and the one freshly combed, requires precise setting and timing. Especially, it is related to the timing of detaching roller forward motion.

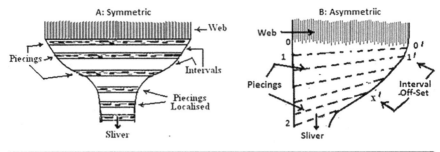

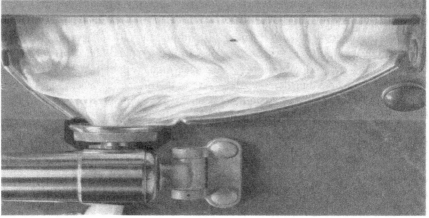

FIGURE 3.51   A: Symmetric and B: Asymmetric condensation:[2,5] In conventional comber the web condensation was done by placing the trumpet centrally in front of detaching rollers. On all modern combers, it is slightly displaced. This minimizes variation due to piecing wave.

intervals along its length. Such a sliver when tested on an Uster electronic tester for its evenness shows a marked peak on the graph, corresponding to each localized joint. With asymmetric condensation (Figure 3.51 B and C), the trumpet in front of detaching rollers is displaced from its central position to one of the sides.

Accordingly, the piecing waves are also offset and are spread more along the length than are localized. Thus, the initial joint 1.1' (Figure 3.51 B) transforms itself into 4.4' at a point close to the trumpet. This, in fact, enables the reduction of the amplitude of the piecing wave.

It should, however, be noted that the level of inherent irregularity in either symmetric or asymmetric condensation differs only in the height of such peaks. The peaks become far less significant. However, once the irregularity is introduced, it cannot be totally eliminated.

The graph (Figure 3.52) is taken from an electronic tester and shows the difference in the peak heights. Being periodic in nature, the peaks appear at regular intervals. It can be observed that the peak height with asymmetric condensation is much subdued. It will be also interesting to note that the intervals (A and B) indicate the distance over which the peaks appear and this corresponds to the distance between two adjoining piecing joints.

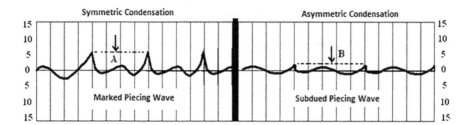

FIGURE 3.52 Uster recorder chart – comber sliver:[3,5] The graph clearly brings out the difference in terms of the reduction in the variability peaks when asymmetric condensation is used.

As the web passes through the trumpet, the table-calendar roller draws the condensed sliver ahead. The slivers coming from individual heads are then laid on the sliver table which extends right across the total number of heads. Depending upon the number of heads (in modern eight slivers and in old six slivers), an equal number of slivers emerge from corresponding table-calendar rollers and carry on their journey along the length of the sliver table. The table surface is made especially smooth and kept polished so that there is absolutely minimum resistance offered by the tabletop for the sliding of slivers. Even then, sometimes the slivers break on the table, only because of their weakness. Negligence on the part of the tenter in either quickly noticing these breaks or piecing them badly can still lead to irregularity in the final delivered sliver.

In some instances, when the sliver from any head is likely to be missed for a longer duration, either due to lap-ups or exhaustion of lap, the tenter while running the machine follows a practice of using some extra length of sliver from the reserve can in which he normally stores a sliver waste. The missing length of slivers owing to the above reasons (lap-ups, etc.) is made up by the sliver from the reserve can. The tenter is expected to take care in doing this practice. Even when such a practice saves some comber sliver waste, it should not be encouraged at any time, as it mainly depends upon the judgment of the person doing it. It is a better practice to educate the worker to avoid the production of comber soft waste (good sliver waste) rather than misusing it. Finally, all the slivers are combined and directed towards the draw box for drafting purposes.

It is very important to see that stretching or any kind of excessive tension is avoided after the sliver formation stage. Sometimes, French chalk powder is used to make the surface smooth. Similarly, the drafts between the table-calendar roller and the detaching roller and that between the back drafting roller and table-calendar roller has to be kept very low. In fact, these are the danger regions where even a slight increase in the tension draft can lead to disastrous consequences. A tension draft should never exceed 1.05.

3.1.12 DRAW BOX AND COILING

The slivers coming from the sliver table are assembled and made to occupy a width which is suitable for drafting rollers (Figure 3.53). In conventional combers, a four-over-four drafting system was used and the draft employed was almost equal to the number of heads on a machine.

With earlier versions of the modern comber, two-over-two single-zone drafting was introduced with a draft around three in the draw box.

Sliver Drafting & Web Condensation

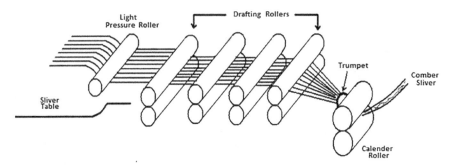

FIGURE 3.53 Drafting and web condensation (conventional):[1,2] The draw box combined slivers from each head. The draft employed decided the hank of the sliver delivered. After passing through the trumpet, the sliver was finally coiled into comber cans.

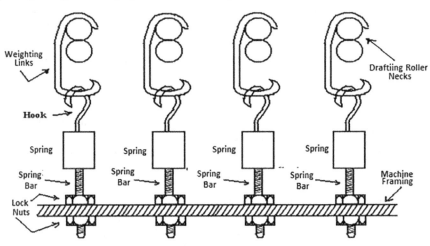

FIGURE 3.54 Spring loading on drafting rollers:[1,2] The coiled spring with weighting hooks was used to load the drafting rollers. The links connected from the spring were hung on top of top roller bearing blocks.

This resulted in a delivery of a thick sliver. It also helped in better handling of comber sliver (comparatively less stretching). The spring loading system on drafting rollers (Figure 3.54) was used in earlier models of combers. This type of roller loading was used on earlier versions of the then-modern combers. Those combers had a provision of processing slivers in two groups of four each.

The drafting system then delivered two slivers thick enough and they were either twin-coiled or bi-coiled[*]. In such cases, the comber had either one or two delivery cans respectively. The facilities were also provided for changing the draft on these machines.

---

[*] Two slivers either coupled or put separately into the same can.

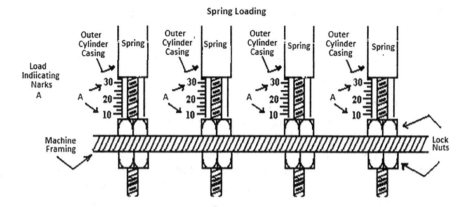

FIGURE 3.55 Spring loading on drafting system:[2,5] The compressed springs encased in steel cylinders offered very precise pressure on top drafting rollers. The compression of the springs can be adjusted against the markings engraved to vary the pressure.

Again, as against the traditional lever weighting system, all earlier versions of modern combers had a spring loading system (Figure 3.55) where the springs were encased into a compact cylinder and the markings on it showed the corresponding weighting, The provision was made to adjust the position of the lock-nuts so that the spring pressure within limits could be varied (Figure 3.55). The typical bouncing-off defect associated with earlier hanging dead-weight system was thus totally eliminated with spring loading system. However, it becomes essential to check the calibration of the springs after a certain period of working.

3.1.12.1 Phasing of Piecing Wave and Two-over-Two Drafting System[1,3]

Peculiar drafting conditions exist at the comber draw box, owing to periodic irregularities in the sliver entering the drafting system. When the material passes through any drafting system, there is an apparent increase in the diameter of the bottom roller according to sliver thickness. As this sliver thickness is likely to vary at every joint of the piecing wave, the slivers when combined in the drafting system, are drawn at different rates.

With several pairs of drafting rollers arranged in a series, a selective action takes place at each roller line. This tends to bring a periodic variation of individual slivers into phase (all the thin or thick places in the sliver appearing together in the final drawn-out sliver). This results in the periodic variation of large amplitude in the final sliver. It has been established that the wavelength of this variation bears a direct relationship with the piecing wave in the individual head sliver and the draft employed in the draw box. The amplitude of the draw-box sliver wave varies directly with the quality of piecing. The draw-box sliver may also contain long-term irregularities due to uneven laps made on conventional lap preparation.

However, the most significant factor affecting regularity is piecing itself. Unless good piecing is obtained, a periodic irregularity of large amplitude always persists, especially when a multi-roller draw box is used. The wavelength of this irregularity is:

$$\text{Wave Length} = \text{Length of Piecing} \times \text{Draft Employed} \qquad (3.1)$$

3.1.12.2 Comparative Performance of Drafting Systems

Thus, the inherent periodic variation of the same wavelength of each of the comber head slivers, are not totally reduced by the process of doubling at the draw box. In fact, owing to phasing, they get further modified. Experimentally also, it has been proved that, not only the periodicity of the comber sliver is related (as given by expression (3.1) above) to the draft employed in comber but also, the subsequent drawing operations increase this wavelength in proportion to the draft employed at those stages.

In a conventional comber draw box with four or five pairs of top and bottom rollers, this phasing tendency is seen to be marked. When the number of pairs are reduced to two (2/2 drafting), there has been a substantial reduction in the variation in the comber draw-box sliver. This is because in conventional combers when each pair contributes towards the ultimate alignment of the piecing wave, with single-zone drafting (2/2 system), there is a minimum opportunity provided for the piecing joints to move into phase. (Figure 3.56)

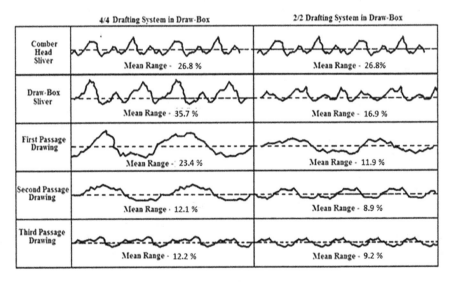

FIGURE 3.56 Comparative performance: 4/4 versus 2/2 drafting – extended research revealed that the use of a conventional drafting system involving four pairs of drafting rollers often led to the phasing of the piecing wave. Reducing the number of drafting zones was one major solution to reduce the phasing effect of the piecing wave (at constant waste extraction level).[3]

Both 2/2 and conventional drafting systems on combers are shown in photographs (Figure 3.57 (a) and (b)). Older versions of modern combers preferred to have a single-zone drafting system in their comber draw box. Moreover, it was also found that, in addition to improving short-term irregularity, this system at the drawing frame provided a minimum level of irregularity at the second post-comb draw frame passage.

Though the 2/2 drafting system was primarily designed for the comber draw box, it is equally effective for subsequent drafting operations with the comber sliver. It is interesting to note that (Figure 3.58), there is a significant improvement in the mean range percentage after the first passage of drawing. Further, after the second passage of the draw frame, the difference in the mean range between 4/4 and 2/2 is considerably reduced.

However, the third passage of draw frame does not seem to contribute much. It will also be noticed that 2/2 drafting keeps the mean range values at a noticeably lower level.

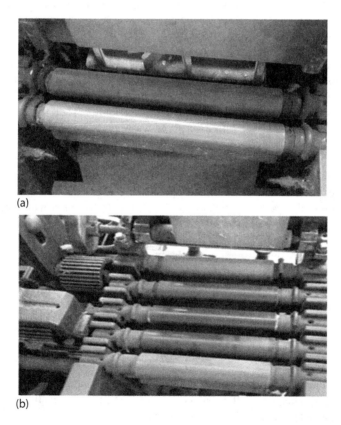

(a)

(b)

FIGURE 3.57   2/2 and 4/4 drafting system: A: Two-over-two draw box B: Conventional draw box. (Courtesy from CIRCOT – Photographs taken in CIRCOT workshop by author)

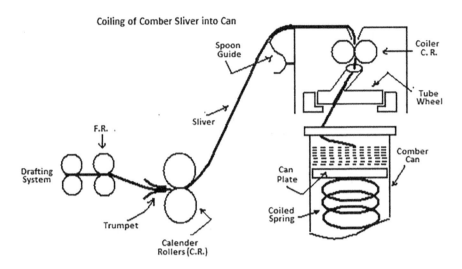

FIGURE 3.58 Coiling at the comber:[1,2] Like any other sliver material, the comber sliver is also coiled into cans. The sliver being soft and weak, much care is required to be taken in avoiding its stretching or breaking.

From fig. 3.56, it can be again observed that it is only after the second drawing passage (post comb) that the sliver variation is significantly reduced. So also, after the second passage, the mean range percentage levels of sliver from 2/2 drafting is noticeably lower. The comber draw-box sliver deteriorates when the waste percentage is increased. In general, with a very low waste percentage, the performance of both 4/4 and 2/2 is inferior.

The regularity is influenced by the draft used in the draw box with the optimum range depending upon the number of slivers fed, the capacity of the drafting system and whether there is single- or bi-coiling. While using a 2/2 single-zone drafting system, there is always a limitation on the amount of drafts that can be used. Thus, a draft of not more than 3.0 is suitable for satisfactory working.

However, it was found that, with lighter laps (35 g/m or 500 grs/yard) and for finer mixing, the 2/2 drafting system gives the best results. In some of the mills, when finer mixings were worked on slow-speed combers (conventional), modifications in drafting from 4/4 to 2/2 gave better results. In modern combers, however, the drafting capacity of the draw box is very much improved and better results are obtained even with the heavier laps (see Chapter 9).

3.1.13 Coiling

The sliver from the draw box is passed through a trumpet having a still smaller aperture of the hole. The sliver is then calendered, so as to further consolidate the drafted material. The coiling resembles exactly the one on the card or draw frame.

As the comber sliver is considerably weak, it is necessary to keep an absolutely minimum tension draft of 1.02 to 1.04 in between (a) calendar roller and front roller and (b) coiler calendar roller and calender roller. This is very important to avoid any stretching of slivers. In some of the combers, the hanging length of the sliver is reduced by suitably placing the calender rollers and coiler closer to the front of draw box. In conventional comber, the can size and therefore the can capacity was so chosen as to accommodate almost one full lap-weight worth of material into the can.

With conventional combers, the common can sizes were – 9″ × 36″ and 12″ × 36″ (diameter × height). With modern machines, the can size steadily increased – 14″ × 36″, 16″ × 36″ or 22″ × 42″.With heavy cans, it becomes necessary to provide casters under the cans. This facilitated easy transportation of the heavy cans from the comber section to the post-comb drawing section.

REFERENCES

1. *Manual of Cotton Spinning: "Draw Frames, Comber & Speed Frames": Frank Charnley*, The Textile Institute Manchester, Butterworths, 1964
2. *Elements of Cotton Spinning: Combing: Dr. A.R.Khare*, Sai Publication
3. New techniques in Combing: For greater regularity of combed sliver, *Platt's Bulletin*, Vol.8, No.11
4. *Spun Yarn Technology – Eric Oxtoby, Sen. Lecturer in Yarn Manufacture, Leicester Polytechnic, U.K.* Butterworth Publication, 1987
5. *Manual of Textile Technology: "Technology of Short Staple Spinning": W. Klein*, The Textile Institute, Manchester.
6. Platt's Century comber booklet
7. Whitin Comber booklet
8. Rieter's Comber E 7/4 Comber pamphlet & booklet

4 Important Comber Settings

4.1 IMPORTANT COMBER SETTINGS

As the main function of the comber is to extract short fibres, there is more than one setting that governs or controls the waste percentage and some of these settings are often used to vary the comber noil. In fact, these settings can be classified into two categories – in the first, there are those settings which are changed to vary the comber noil; whereas; in the second, the noil% is affected when the settings are altered.

Whenever the noil% of the comber is to be changed for bringing the necessary changes in the quality of the output, the following settings are altered: (1) the setting between back bottom steel detaching roller and bottom nipper (called 'step gauge'); (2) the setting between bottom nipper and cylinder half lap (leaf gauge) and (3) the setting of top comb – its timing, penetration, distance from back bottom detaching roller and its angle of penetration with respect to vertical.

4.1.1 SETTING BETWEEN BOTTOM NIPPER AND STEEL DETACHING ROLLERS

It is already known that the nippers have forward and backward swinging motions. During backward motion, combing by cylinder needles is carried out; whereas in forward motion, the piecing-up operation takes place. In conventional combers, the nippers are closest to detaching rollers at index 19. The nippers move very slowly around this index to facilitate better piecing. In fact, between index 18.75 and 19.25, the nippers move so slowly that they appear to be almost stationary. As explained earlier, index 19 is, therefore, called the 'dead centre of the comber.

The distance between the bottom nipper and back bottom detaching roller is called the detachment setting. It controls the extent to which the freshly combed fibres are brought closer to detaching roller nip. Thus, the closer this distance, the easier it becomes for the combed fibre to reach detaching roller nip. As a result, the proportion of the fibres which is restricted by the top comb is correspondingly less. Ultimately, less noil is extracted.

The setting procedure is very simple and easy. Initially, the index wheel is turned to number 19 and the distance between the bottom nipper and back bottom detaching roller is adjusted by using the two lock nuts – labelled A and B in Figure 4.1 and with the help of a step gauge (different gauges from 8/32" to 16/32" are arranged in steps on different faces of this gauge and hence the name step gauge) held in between the two as shown in Figure 4.1).

It can be seen that the minimum gauge, in this case, is 8/32" below which, though it may be possible to set the distance (not with the step gauge), it may pose problems

DOI: 10.1201/9780429486555-4

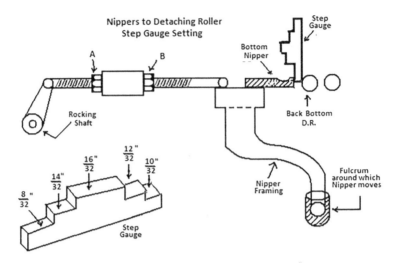

FIGURE 4.1   Step gauge:[1,2] The most important gauge that decides the level of waste extraction level in the comber. When the gauge is widened, the waste increases and vice-versa.

and would not permit the entry of the top comb. As there are six heads (in modern combers eight heads), it is required to do this setting on all the remaining heads. Many a time, when the comber is taken for regular periodic cleaning or for full overhauling, it is a common practice to carry out this setting procedure on all heads before starting the comber for regular production..

Equally important is to see that when nippers are set with a detaching roller, the two ends of the nipper are at an equal distance from it. A person carrying out the step gauge setting, therefore, has to move the gauge across the nippers, to and fro, to feel this same distance. This feel is developed only after some experience. The gauge during its movement across the nippers has to be held firmly between the thumb and first finger. So also, the pressure which is required to move the gauge across has to be uniformly felt. Further, it is very important to hold the gauge absolutely vertically (Figure 4.1) so that its two faces touching either side (bottom nipper on one side and the detaching roller on the other side) correctly measure the distance marked on the face of the gauge.

4.2   QUADRANT SETTING (MASTER GAUGE)[2,5]

In the normal working of the comber, the detachment setting is changed only when the comber noil is required to be changed or else when the cotton with a totally different effective length is to be processed. However, changing the step gauge on every head is a time-consuming process and would lead to a loss of production time.

For this purpose, a quadrant setting or master gauge is provided on the machine (Figure 4.2). When altered, this setting changes the step gauge on all the heads simultaneously. In fact, for this, Tightening Bolt A which fixes the position of the quadrant has to be loosened first. Simultaneously, Retaining Bolt B on each individual head is also required to be loosened. The quadrant can then be lifted up or pushed down so that the number mark on the quadrant against the pointer is either raised or lowered.

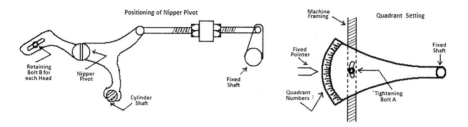

FIGURE 4.2 Quadrant or master gauge:[1,2] It is called a 'master gauge' as it controls the movement of the nipper stud, which ultimately has a similar effect of changing the step gauge.

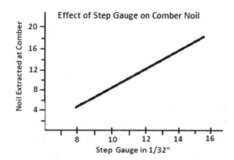

FIGURE 4.3 Varying step gauge:[1,2,4] As the step gauge directly controls the distance between the nippers and the detaching rollers, it has a profound influence on the waste extracted at comber.

With a higher mark on the quadrant, coming against the pointer, the detachment settings (between the bottom nipper and detaching roller) on all the heads are simultaneously increased.

It is a common experience that when the quadrant number increases by 1, the waste extracted by the comber increases by approximately 2%. The two graphs (Figures 4.3 and 4.4) reveal that the effect of varying the step gauge or quadrant setting on comber noil is almost similar.

It is found that when the comber noil is increased by widening the step gauge, the performance of the comber in extracting the short fibres below a certain predetermined level is enhanced. This ability of the comber to fractionate short fibres from long fibres is called as its fractionating efficiency (see Section 6.1.4.6).

4.3 POSITIONING OF NIPPER PIVOT STUD[2,5]

The setting of the nipper pivot stud is done at any index. In this, the distance of the nipper stud is adjusted with respect to detaching rollers by a bowl and brass gauge (Figure 4.5). The bowls with different thicknesses (1/4", 3/8", 1/2" etc.) are available

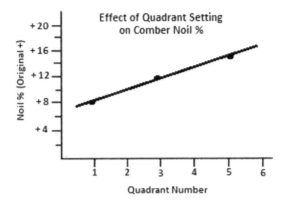

FIGURE 4.4 Varying quadrant setting:[1,2,4] When the quadrant setting is changed, it affects the positioning of the nipper pivot stud. It has a similar effect of changing the step gauge.

and depending upon the need, any one of them can be chosen. With a larger diameter, the nipper pivot is shifted farther away from the detaching rollers, thus leading to higher comber noil.

Once the nipper stud is fixed with the gauges shown in Figure 4.5, its position remains the same during the combing cycle irrespective of index numbers. Hence, the setting is done at any index. The stud can be adjusted by putting a bowl of the desired thickness. Retaining Bolt A is then loosened. Then with the help of two check-nuts, B and C, the appropriate position of the bowl around the stud can be fixed by using brass a gauge.

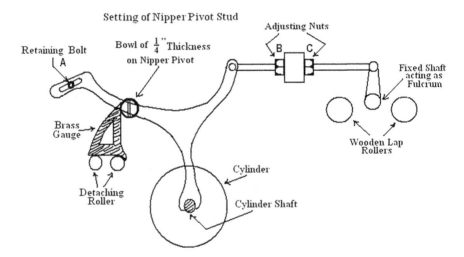

FIGURE 4.5 Bowl gauge:[1,2] It fixes the position of the nipper pivot stud around which the nippers swing forward and backward. The setting indirectly controls the distance between detaching rollers and nippers.

It can be seen that the positioning of the nipper stud with respect to the detaching roller, actually decides the distance between the detaching rollers and the nipper, when the latter is closest to the former at index 19. Obviously, when the bowl with a larger thickness is used, the nipper stud is thrown relatively a little back and away from the detaching roller. This is why comber noil (%) increases. The bowl with a smaller thickness is used for lower-grade cottons whereas the one with a larger thickness may be used for finer varieties which need higher noil extraction.

Another interesting and important point is that the shifting nipper pivot stud also changes the position of the archway movement of the nippers over the cylinder needles. Therefore, after the bowl gauge setting is carried out, it becomes necessary to again check the two important settings – (1) setting between cylinder needles and bottom nipper and (2) step gauge setting.

4.4 SETTING BETWEEN BOTTOM NIPPER AND CYLINDER HALF LAP[1,2]

The action of the half-lap needling is based on a progressive increase in the severity of the combing action. As mentioned earlier, the first few rows of needles while passing through the lap fringe initially prepare the fleece. The subsequent rows of needles, after this initial preparation, carry out a thorough combing. In accordance with this, the needles of the first few rows are a little widely spaced and slightly shorter in height. Subsequently, the height of the remaining rows remains constant. Therefore, this setting is very important as it governs the effective penetration of cylinder needles. Owing to progressive penetration, the closest setting between half-lap needling and nippers is thus done only when the needle height becomes constant. This happens to be at or around index 30 (conventional comber).

Thus, depending upon the needling system this setting is carried out when the needle height becomes constant. For moving the nippers up or down to increase or decrease the distance respectively (Figure 4.6), Fixing Bolt C on either side of

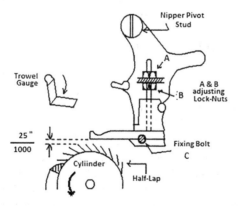

FIGURE 4.6 Setting between bottom nipper and cylinder half lap:[1,2] This decides the depth of penetration of cylinder needles into the lap fringe at the time of combing by cylinder needle.

the head needs to be loosened. This helps in unlocking the nippers from the nipper frame assembly and makes the movement of the bottom nipper free.

The actually required movement of the bottom nipper in either direction is brought about by the combined movement of adjusting nuts (A and B). A special gauge called 'trowel gauge' is used to adjust the distance between the needles and the bottom nipper. Due care, however, should be taken whenever the nippers are to be lowered. In this case, any slackness between the two nuts (A and B) proves to be detrimental. Therefore, while carrying out the setting, the movement of the nippers should be so chosen that the nippers are finally made to move in the upward direction (i.e., by tightening nut A). Further, after the setting is complete, it is always necessary to check two things: (1) full tightening of Fixing Bolt C so as to secure the nippers firmly to the nipper framing and (2) to examine and check the step gauge setting. The checking of the step gauge becomes necessary as the position of the bottom nipper changes in a vertical plane and hence change in the nipper movement in the vertical direction is likely to affect the detachment setting.

It is possible to change the distance between the bottom nipper and cylinder needles by selecting an appropriate trowel gauge. However, as this setting alters the penetration of the cylinder needles, it is rarely changed when the same mixing is processed. The setting may be required to be changed only when the lap weight (which changes the lap thickness) is changed.

4.5 TOP COMB SETTING

The top comb influences the amount of waste extracted in four different ways: (a) its timing of entering the fleece, (b) its depth of penetration into the fleece, (c) its distance from detaching rollers and (d) its angle of penetration into the fringe.

4.5.1 ENTRY TIMING

The normal timing for the top comb to enter the fringe is after the piecing of two fringes. Thus, when the piecing of the main fringe is being done and later when the pieced-up joint comes under the detaching rollers around index, say 12 to 12.5, the earliest that the top comb can enter the fringe would be not before index 13. However, this can top comb can enter the fringe would be not before index 13. However, this can also be delayed as late as index 13.5 or proponed at an earlier index, around 12.75. The normal entry of the top comb usually takes place at index 13. It must be remembered that though the first few longer fibres reach the tailing ends of the already combed fringe around index 8 or 9, the detachment of the main fringe starts later after index 12. Hence, the top comb should be allowed to enter the fringe only after this index. For adjusting the entry of the top comb (Figure 4.7), the desired index (say index 13) on the index wheel is brought against the pointer. With the help of lock nuts A and B, the position of Lifter Bracket N is adjusted so that there is a paper gauge (very small gap) between the platform of Lifter Bracket N and the bottom face of the bowl. With lock nut B, the extreme position of the bowl at the edge of the platform can be

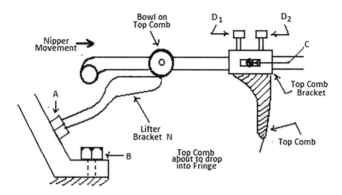

FIGURE 4.7 Top comb entry timing:[1,2] The top is withheld and not allowed to enter the fringe till the pieced-up joint comes under the nip of back detaching rollers.

adjusted. Therefore, when the nippers move further, the bowl is no more held by the platform and thus drops down into the fringe.

It may be noted that the entry timing of the top comb has a greater influence at a closer step gauge. Thus, when a smaller step gauge is used, early or late entry of the top comb can cause significant changes in the noil extracted.

4.5.2 Top Comb Penetration

The depth of the top comb tip (Figure 4.8 (a)) with respect to the top surface of detaching roller governs the top comb penetration into the fibre fringe. As shown in this figure, it is the depth A, below the plane of the fibre fringe. The top comb is said to be 'level' or 'zero-setting' or 'flat' when the tips of its needles are in line with the top surface of detaching rollers. To adjust such positions, the setting is done with D_2 (Figure 4.8 (b)).

The top comb can be lowered below this line so that it goes a little deep into the fringe (e.g., positive penetration 1/16"). As against this, when the top comb is raised above this level (e.g., negative penetration 1/16"), it would mean that the tips of top comb needles are raised above this horizontal plane.

The effectiveness of the top comb action is also related to the position of the bottom nippers (or cushion plates). As shown (Figure 4.9 (a)), if point A is below the top surface of the detaching rollers (point P), then for the same depth of top comb, its influence on the fringe being detached will be reduced and vice-versa.

In general, the deeper the top comb is set into the lap fringe (Figure 4.9 (c)), the more the amount of waste extracted. Similarly, when the top comb is lifted up (Figure 4.9 (b)), the waste extracted reduces. This is because; the influence of the top comb in restricting those fibres which would otherwise go into form the sliver, is reduced.

The required position of the top comb (its depth), as mentioned earlier, is obtained by adjusting the screw D_2 (Figure 4.8 (b)). It must also be also remembered that apart

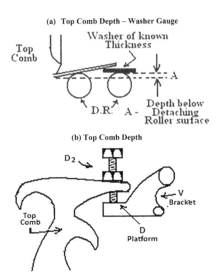

FIGURE 4.8 (a) and (b) Depth of the top comb:[1,2] The depth of the top comb is always expressed with reference to the plane which includes the top surface of the bottom detaching rollers. If the needle tips lie in this plane, it is termed as 'top comb in level'.

from restricting the fibres that have not reached the detaching rollers at the time of top comb entry, the presence of top comb needles in the fringe causes interference in the free flow of the detached fringe. Too much interference, merely to extract more waste, makes the web lose its smooth transparent form and appears to be torn in places.

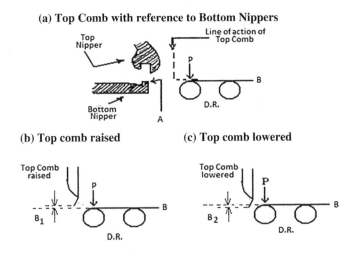

FIGURE 4.9 (a), (b) and (c) Relative placements of top comb and bottom nipper: The top comb can be set a little deeper or can be lifted up w.r.t. plane B; to intensify or reduce its action. The position of the top surface of the bottom nipper matters in this case.

4.5.3 DISTANCE OF THE TOP COMB FROM DETACHING ROLLERS

The position of top comb needles with respect to their distance from the detaching roller is another important criterion. It decides the closeness of the needles from the detaching line and, therefore, controls the length over which the top comb can effectively influence the short fibre restriction (Figure 4.10 A).

In other words, it controls the extent to which the top comb can exercise its influence on short fibres, which otherwise may go into sliver. A closer setting allows the top comb to exercise more control, leaving a wider distance Z (Figure 4.10 B) of lap fringe behind it. This obviously leads to more comber noil.

Conversely, with a wider setting, the distance X becomes wider, thus shortening the distance Z and reducing the influence of the top comb over a fringe. As a result, it reduces the waste extracted.

The actual setting is done by shifting the top comb holding bracket (Figure 4.12 A and B) in either direction by loosening Lock Nut N. However, it is essential to choose this setting very carefully. Too close a setting is likely to increase the possibility of top needles touching the detaching roller at times. This, in a way, causes harm to the needles or else causes damage to the back top detaching roller. In some cases, it may also lead to more interference of the top comb which is very likely to restrict the smooth detachment of fibres. This can be easily noticed from the web appearance which appears to be disturbed (cuts in the web). Too wide a setting, on the other hand, leads to poor fractionation. As shown in Figure 4.11, the required adjustment can be made (another make) by adjusting the two nuts P_1 and P_2. By tightening nut P_2, the top comb moves towards the detaching roller. As mentioned earlier, in another mechanism (Figure 4.12 A and B), the top comb holding bracket itself can be moved towards or away from the detaching roller.

This is done by loosening the adjusting bolt N and then shifting the bracket as desired. In a Whitin comber, the top comb is carried in the slotted bracket which is

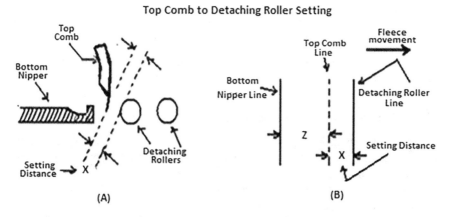

FIGURE 4.10 [A] and [B] Distance of top comb from the detaching roller:[1,2] It controls the influence of the top comb over the length of the fringe remaining behind the top comb needle line.

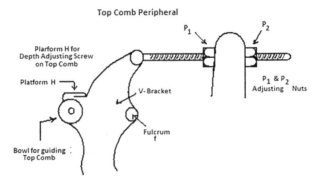

FIGURE 4.11   Distance of top comb from D.R.:[1,2] When the top comb is closer to detaching rollers, it is able to influence over a wider length of the fringe remaining behind its needle line and hence the short fibres are more effectively extracted.

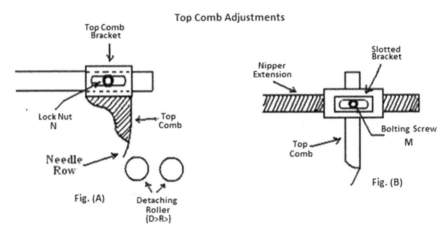

FIGURE 4.12   Top comb holding bracket:[2] In some typical combers a simple arrangement is made to move the top comb bracket towards or away from detaching rollers.

fixed on the extension of the nippers (Figure 4.12 B). By loosening Bolting Screw M, the slotted bracket can be shifted to get the required setting.

4.5.4   Angle of Top Comb Penetration

The angle with which the top comb penetrates the lap fringe can be adjusted by the 'angle gauge' (Figure 4.13 a, b and c). In fact, the process can be compared with the combing of human hair where the people use easier angle while initiating combing their hair. In that, when the comb makes the acute angle with the direction of its motion, the action is comparatively lighter. However, on a combing machine, the comb is stationary; whereas, the fringe moves ahead. In this case, one can imagine the top comb having a relative motion in the backward direction (choosing the frame of reference of the fringe to be stationary).

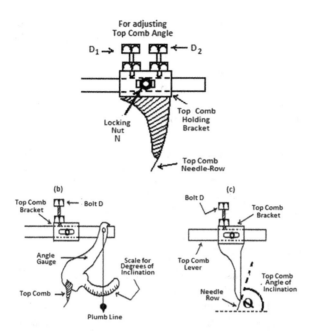

FIGURE 4.13 (a) Top comb angle adjustment:[1,3] In typical cases, it may be necessary to change the top comb angle so as to vary its action on the lap fringe being detached. Top comb angle of penetration:[1,2] The actual angle can be varied through a small margin. This angle of penetration greatly controls the severity of top comb action – whether harsher or softer.

Thus, as indicated in Figure 4.14, if line AB is the line of direction of the fringe movement, the top action will be less severe when its inclination is $P_3Q_3.B$.

With an acute angle P_1Q_1B therefore, the influence of the top comb action will be more severe. In the figure, the angles of the top comb inclinations are a little exaggerated to illustrate their impact. In actual practice, these angles are comparatively very small. Even then, the effect of the inclination can still be perceived.

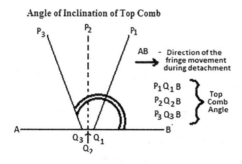

FIGURE 4.14 Angle of top comb:[1,2] The angle of inclination of the top comb is always measured with reference to the direction of the motion of the lap fringe.

The necessary arrangement (Figure 4.13 b and c) can be made by doing the adjustment with Bolt D. The special angle gauge with a plumb line attachment is used to choose the angle of inclination of the needles. As shown in Figure 4.13 a, the change in the angle can also be brought about by using either of the two nut bolts – D1 or D2. Also shown in Figure 4.12 (A) and (B) is Lock Nut N, which, as explained earlier can be used to move the whole bracket towards or away from detaching rollers. to respectively increase or decrease the top comb influence.

4.6 OTHER SETTINGS ON COMBER AFFECTING COMBER NOIL

Other than the above settings which are used to intentionally alter the comber noil, there are a few other settings, which when altered can affect the noil extracted. Some of them have direct influence whereas; some indirectly alter the noil extracted.

4.6.1 BRUSH SHAFT SETTING

The very purpose of the brush shaft carrying the brushes is to ascertain that the cylinder needles are very effectively stripped in every combing cycle. It is evident, therefore, that its setting with the needles does affect the noil removed by the brush. Thus, if the brush bristles do not penetrate the needles sufficiently, the cleaning of the needles is seriously affected. This inadvertently affects the waste removed by the brush.

Further repercussions of these are that the cylinder needles would steadily get loaded with fibres. This would again affect each subsequent performance of needles in removing short fibres from the lap fringe presented to them. The brush shaft setting, therefore, indirectly affects the comber noil. However, as stated earlier, this setting is never to be changed for varying the comber noil. The setting is thus changed only when the bristles are worn out and fail to adequately penetrate the needles. In this case, it is the brush shaft which is required to be raised.

The normal procedure is to adjust the penetration of the bristles (Figure 4.15) on one single head at one of the ends. The person carrying out this setting has to ascertain that the distance does give a certain level of penetration of brush bristles. Using the gauge (which is adjustable), this distance is marked on the gauge and then, the same distance is adjusted on the remaining heads. After the setting is carried out there are certain things which need to be observed. First, the brush shaft has to move comparatively freely with the level of penetration of the bristles. Second, one has to make sure that when the machine is run with this setting, the cylinder needles are completely stripped on all the heads. Third, it must be remembered that too deep a setting is going to cause serious damage to needles as well as brush bristles.

4.6.2 TYPE OF FEED, FEED TIMING AND LENGTH FED[1,3]

4.6.2.1 Forward and Backward Feed

During the combing cycle, the feeding of the fresh lap fringe can be done when either the nippers are moving in the forward direction or when they are receding back. It is interesting to note that during the forward movement of the nippers, they

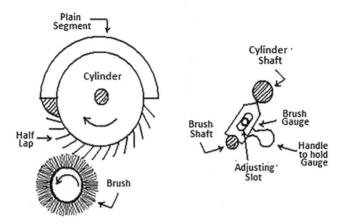

FIGURE 4.15 Brush shaft setting – gauge:[1,2] The brush bristles are required to penetrate only a little the cylinder needle. Too much penetration is more likely to cause damage to the needles.

are open and hence, when the feeding takes place during this period, the projecting fringe, approaching detaching rollers, advances further by a distance equal to the feed length. This is, assuming that the nippers start opening a little later than the timing of the feed roller. However, with backward feeding, the lap is fed during the backward motion of the nippers. During this period, the nippers, while receding also close and hold the lap fringe firmly for the ensuing combing operation by cylinder needles. Thus, a comparatively shorter time is available for the feeding to complete its action before the main combing by the cylinder. In practice, only a part of the feeding takes place before the nippers are fully closed. With the help of the Baer sorter diagram (Figure 4.16) which shows important fibre characteristics, the principle involved can easily be explained.

Let D and f represent the detachment setting and feed length per combing cycle respectively.

Let it be also assumed (Figure 4.17) that:

$$S_2S_2' = D; \quad S_1S_1' = D + f; \text{ and } S_3S_3' = D - f$$

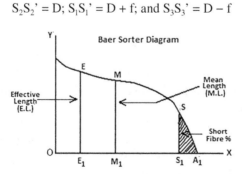

FIGURE 4.16 Baer sorter diagram:[1,2] Important fibre parameters such effective length, mean length and short fibre% are depicted by Baer sorter diagram

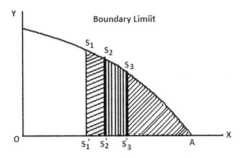

FIGURE 4.17 Influence of detachment setting and feed Llength on the comber noil:[1,2] During detachment of the fringe, all fibres shorter than the detachment setting and those stopped by the top comb are expected to be removed by the cylinder in the next cycle.

Thus, in the forward feed, all the fibres shorter than 'D' are expected to go to waste. With f as the feed length during detachment therefore:

[Fibres < (D – f)] will go into comber noil and

[Fibres > D] will form the comber sliver.

Actually, the fibres in between these two limits would either go into noil or sliver depending upon their disposition at the time of detachment. Thus, the boundary of fibre length demarking the areas of the fibres going into waste and forming sliver can be safely assumed to be halfway between the above two limits.

$$[D + (D – f)] / 2 = [D – f/2] \tag{4.1}$$

In the backwards feed, the length of the fringe of fibres presented to the cylinder is (D + f). This is because; D is the length of the fringe projecting in front of the nippers after the detachment and f is the length fed before the nippers close for combing by cylinder, assuming that all the length fed advances in front of nippers (though it is not the case). Hence:

[Fibres < D + f] will be removed as noil

After the completion of combing by cylinder in backward feed, the nippers move ahead and are closest to detaching rollers at some later index number. Here, the fibres shorter than detachment setting (D) will all go to noil.

So, [Fibres > D] will form the comber sliver.

Again, the fibres in between these limits would either go into waste or form the sliver, depending upon their actual disposition in the lap fed. Thus, the boundary length demarking the area here would be as follows:

$$[D + (D + f)] / 2 = [D + f/2] \tag{4.2}$$

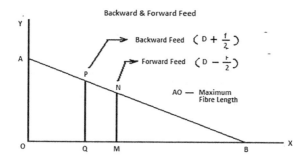

FIGURE 4.18 Boundary length:[1,2] It is an assumed length below which all the fibres are expected to be removed by the cylinder.

From Equations (4.1) and (4.2) above we can represent these lengths in a triangle as shown (Figure 4.18)

$$PQ = D + \tfrac{1}{2} \text{ Backward feed length i.e. } (D + 0.5 \text{ f})$$

$$MN = D - 0.5 \text{ Forward feed length i.e. } (D - 0.5 \text{ f})$$

$$AO = \text{Maximum fibre length}$$

Suppose the Baer sorter diagram is drawn in such a way that its area is equal to the area OAB.

Further, PQ and MN represent the areas of demarcation of fibres in sliver and in noil in Backward Feed and Forward Feed respectively then:

In Backward Feed

% Waste Extracted

$$= \left[\Delta BPQ / \Delta BAO \right] \times 100 \qquad (4.3)$$
$$= \left[PQ / AO \right] \times 100$$

In Forward Feed

% Waste Extracted

$$= \left[\Delta BMN / \Delta BAO \right] \times 100 \qquad (4.4)$$
$$= \left[MN / AO \right] \times 100$$

From the expressions (4.3) and (4.4), one can easily see that the value of (4.3) will always be greater than that in (4.4). This means that, with all other settings remaining constant, the waste extracted in the backward feed will be more than that in forward feed. Experimentally also this can be proved. By taking an example, it can also be explained.

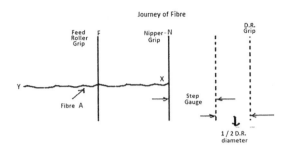

FIGURE 4.19 Gripping line for fibre:[1,2] Though the amount of feed in every cycle remains the same, the journey time of fibre to reach the detaching line may differ depending upon the mode of feed.

To start with, it is assumed that the effective length of fibres processed is (say), 24 mm; the detachment setting is 6 mm feed length is 6 mm and the distance between the nipper bite and feed roller nip is 5 mm. Let us also assume that from their position shown in Figure 4.19, the nippers are moving forward. In a typical case, a fibre A of length XY can be assumed to be lying with its leading end at the nipper grip. The movement of the fibre in backward and forward feed in each case with the reference to the combing cycle is shown in Table 4.1.

In forward feed, the fibre gets detached after the fifth cycle by the detaching roller, i.e., it receives four combing cycles by then. As the nippers are moving ahead, during forward movement, the fibre will advance a distance equal to the feed length (6 mm). But its trailing end is still behind the feed roller (18 mm). So, it will not be

TABLE 4.1
Movement of Fibre in Backward and Forward Feed

Combing cycle number starting from cylinder combing	Forward Feed		Backward Feed	
	Point X	Point Y	Point X	Point Y
	w.r.t. line N	w.r.t. line F	w.r.t. line N	w.r.t. line F
	At the time of Detachment		At the time of Detachment	
1st	6 mm	18 mm	0 mm	24 mm
2nd	12 mm	12 mm	6 mm	18 mm
3rd	18 mm	6 mm	12 mm	12 mm
4th	24 mm	0 mm	18 mm	6 mm
5th	---	---	24 mm	0 mm

detached. During the first combing cycle that follows the first detachment, the fibre project in front of the nippers only partially (6 mm).

As against this, in the backward feed, the fibre is fully held behind the nipper line during detachment (no feeding). So, it does not get detached. Just prior to combing, a length equal to 6 mm is fed in backward feeding (assuming that it is done before the nippers close for combing). Hence, it will be seen that in backward feed, the tailing end of the fibre will be out of feed roller nip only after the fifth combing cycle. Thus the fibre is bound to receive an additional combing by the cylinder. Actually, the fibre 'A' represents a group of such fibres and hence an additional combing received by the fibre groups would lead to higher noil extraction in the case of backward feed.

Thus, when all other parameters remain the same, and only the mode of feed (say from forward to backward or vice-versa) is changed, the comber noil% also changes. A slight change in the particulars (fibre length = 25 mm and feed length = 5 mm), brings out a slightly different fibre movement through each combing cycle (Table 4.2). It can be seen that, in this case again, the journey for the fibre in the forward feed is shorter by one combing cycle. The waste extracted with backward feed in this case, owing to an additional combing cycle, leads to higher noil extraction.

4.6.2.2   Feed Length and Timing[1,3]

Taking the above example, when the feed length is increased there is a small decrease in the percentage of combings that the fibres receive. With reference to the above two tables (Table 4.1 and 4.2), when the feed length is increased from 5 mm to 6 mm, the percentage change in the number of combings will be as follows:

$$\text{Reduction} = (6-5) / 6 = 16.66\%$$

Obviously, this reduction in the number of combings has an influence on the noil extracted. Therefore, when the feed length is increased, the number of combings

TABLE 4.2
Movement of Fibre in Backward and Forward Feed

Combing cycle number starting from cylinder combing	Forward Feed		Backward Feed	
	Point X	Point Y	Point X	Point Y
	w.r.t. line N	w.r.t. line F	w.r.t. line N	w.r.t. line F
	At the time of Detachment		At the time of Detachment	
1st	5 mm	20 mm	0 mm	25 mm
2nd	10 mm	15 mm	5 mm	20 mm
3rd	15 mm	10 mm	10 mm	15 mm
4th	20 mm	05 mm	15 mm	10 mm
5th	25 mm	00 mm	20 mm	05 mm
6th	---	---	25 mm	00 mm

(combing cycles) received by the fibre fringe is reduced. As a result, the noil% is reduced.

The feed length is normally changed by changing the number of teeth on the ratchet. The time during which the feed roller feeds the lap fringe also bears importance. This again is because; as the fringe is moved ahead during detachment (say in the case of forward feed), some fibres which are even shorter than detachment setting, also advance and come under the nip of detaching rollers. It may, however, be remembered that when the top comb enters the fringe, the leading end of only those fibres which are in front of the top comb and which have also reached detaching roller nip, are pulled ahead (detached). With reference to this, when the feed roller ratchet is made to click (starts turning) a little late, the influence of the top comb prevents the full advantage of the advancement of the fringe. Thus, even when the combed fringe is advanced, the top comb prevents some more short fibres from reaching detaching rollers. Hence the waste% increases.

4.6.2.3 Feed Roller Distance

This distance is required to be set according to the staple length. The longer staple length demands a wider distance and vice-versa. The distance decides at what position, the fibres should be let free from the feed roller nip. This especially may lead to very important during detachment of the fringe as well as during combing. Thus, the longer fibres, with a shorter distance would be held by the feed roller grip for a longer duration. This exposes the fibre fringe for more combings and may possibly lead to fibre damage.

4.6.3 DETACHMENT TIMING

In conventional combers, detaching rollers were given motion through a quadrant and pinion. The quadrant was made to rock to and fro due to the bowl that it carried. The bowl, in turn, was controlled by a cam. It was possible to alter the position of this cam placed on the cam shaft, a little on either side with reference to an index number. Thus, if the detaching rollers were expected to rotate in their normal forward direction at an index number (say) 8, it was possible to initiate this movement at either 8.5 (late) or 7.5 (early). The timing of this forward motion of detaching roller, however, controls the superimposition of the already combed and freshly combed fringes. In a way, it affects the thickness of the joint. Depending upon the trial and error method (because of the fibre array diagram), the correct level of superimposition can be arrived at, to give precise piecing. Whether the piecing has been correct or not can also be verified by testing the sliver thus produced, on an electronic tester. It would exhibit a distinct piecing wave. (Figure 4.20 (a) and (b)): When there is either over or under superimposition, the peak or the trough in the recorder would be quite distinct. With correct timing, the amplitude of these peaks or troughs is found to be the minimum.

For a normal single overlap of comber piecing, it is required that the piecing be symmetrical about its midpoint and also that the piecing overlap be nearly at the half of the previous piecing. The piecing shown in Figure 4.20 (a) and (b) are ideal piecings.

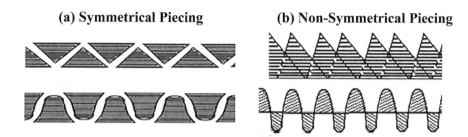

(a) Symmetrical Piecing **(b) Non-Symmetrical Piecing**

FIGURE 4.20 (a) and (b) Type of piecing:[2,3] The ends of the fringe which get superimposed over one another makes a piecing joint. To be an ideal piecing, it has to be symmetrical around its central point.

The data collected by cutting and weighing method at various increments along the length of the piecing joint shows that the most ideal piecing would be A_3 and B_2 (Figure 4.21(a) and (b)) where the weight distribution along the length of pieced-up portion has the minimum variation.

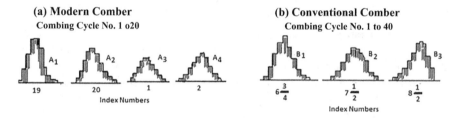

(a) Modern Comber **(b) Conventional Comber**
Combing Cycle No. 1 o20   Combing Cycle No. 1 to 40

FIGURE 4.21 (a) and (b) Effect of timing on detaching roller:[2,3] The timing of starting of the forward motion governs whether the piecing is ideal or not. By slightly varying the timing, one can experiment and arrive at the correct timing to minimise the prominence of piecing wave.

Even the variation in piecing would have its reflections when the sliver thus produced is tested on an electronic irregularity tester (Figure 4.22). Hence by either the cut and weigh method or by obtaining an irregularity curve, it is possible to arrive at the correct detachment timing.

However, as explained before, if the movement of detaching roller is made earlier, it starts taking the fringe a little earlier as compared to the motion of the nippers in the forward direction. This also leaves the top comb to have its influence on a comparatively lesser length of the back portion of the fringe. As a result, it reduces the comber noil.

4.6.4 Lap Weight

It is essential to have better lap preparation, especially when heavier laps are produced for the comber. This is because heavier laps demand better fibre orientation and parallelization; so that the strain on the comber needles and fibre breakage can

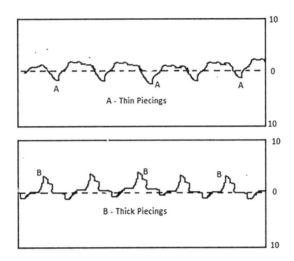

FIGURE 4.22 Irregularity graph for different detaching roller timings:[2,3] This timing basically decides the extent of superimposition. The deviations from the best timing always lead to higher average variation in the comber sliver.

both be considerably reduced. On the other hand, heavier laps also mean thicker material to be held under the nipper bite. In the conventional comber the spring pressure on the top nipper was adequate enough for only lower lap weights (less than 40 g/m). With modern combers, the lap weights are appreciably higher than these levels. Therefore, if heavier laps (>70 g/m) are to be processed, due consideration must be given to the nipper bite; otherwise, there would be an appreciable proportion of good fibre loss owing to fibres slipping under the nipper bite.

The other important aspect when processing heavier laps is insufficient penetration of either cylinder or top comb needles. In both cases, the process of combing becomes incomplete. The satisfactory penetration of cylinder needles ensures better short fibre removal and at the same time, the fibre orientation in the combed material is improved. A poor fibre orientation results in uncontrolled movement of fibres during detachment, leading to poor comber performance. It also puts a heavier burden on the needling. Therefore, heavier comber laps, demand a better lap preparation system.

With a given setting, the waste extracted at comber generally increases with an increase in the lap weight. Here too, the laps prepared with a low number of doublings and lower draft show comparatively more increase when the lap weight is increased. This gets further magnified when longer feed lengths are used. It, therefore, means that any increase in production by increasing lap weight would be successful, only when the fibre parallelization and lap regularity are both adequate.

Achieving adequate lap regularity and better head-to-head comber sliver regularity is also dependent on the type of drafting system used in pre-comb drafting processes. As against the conventional 4/4 drafting, the modern comber lap preparatory systems use 3/3, 2/3, 3/5 or even 4/5. These systems are able to substantially reduce the irregularities in the comber lap, especially those due to roller slip. They also prevent short-term irregularities in card sliver from becoming magnified owing to phasing. However, it must be mentioned here that an additional pre-comb machine (a sequence of three machines) is truly superfluous irrespective of the type of drafting system used. This is because; it does not much improve the lap regularity and also involves additional costs. Further, it leads to the wrong presentation of major hooks. In addition, an additional machine involves higher pre-draft which is very likely to lead to problems such as lap licking at comber.

4.6.5 Cylinder to Steel Detaching Rollers[1,2]

The setting of the plain segment of the cylinder to the back steel detaching roller is standard setting and is not altered. It decides the position of the plain segment on the cylinder. In conventional combers, it is done at index 1.5. For this, it may be necessary to remove the last few needle strips. The bolting screw holding the plain segment and half lap firmly on the cylinder shaft is then loosened. Initially, it is important to see that all the cylinders are correctly centred within each head.

The alignment of the plain segment with respect to detaching rollers is then done using a gauge (Figure 4.23 – the shape of the gauge is magnified for clarity). The reason why this setting and its timing cannot be changed is because this position of both the plain segment and hence the half lap is related to the combing cycle. In fact, it fixes the time at which the cylinder would start combing operation. It may be mentioned here that when this position is reached at an earlier index, the cylinder combing too would start earlier than index 27 (assuming this is the standard index for cylinder combing). However, the occurrence of other motions would have to be at their pre-determined timing. e.g., if the nippers are fully closed by index 26, then the cylinder combing operation cannot start earlier than this index; otherwise, it would lead to the loss of good fibres). The effect of this setting is thus indirect.

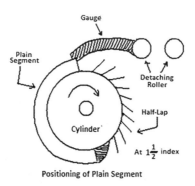

Positioning of Plain Segment

FIGURE 4.23 Positioning plain segment:[1,2] The adjustment of this setting at a particular decides two important events in the combing cycle. One is the start of combing with cylinder needles and the second is the piecing-up operation.

BIBLIOGRAPHY

1. Manual of Cotton Spinning: "Draw Frames, Comber & Speed Frames": Frank Charnley, The Textile Institute Manchester, Butterworths, 1964
2. *Elements of Cotton Spinning: Combing: Dr. A.R.Khare*, Sai Publication
3. *A Practical Guide to Combing & Drawing: W. Klein*, The Textile Institute, Manchester, UK.
4. *Spun Yarn Technology: Eric Oxtoby, Sen. Lecturer in Yarn Manufacture*, Leicester Polytechnic, U.K. Butterworth Publication, 1987
5. Cotton Spinning – William Taggart

5 Cylinder and Top Comb Needling Arrangement

5.1 CYLINDER AND TOP COMB NEEDLING ARRANGEMENT

The needles of both the cylinder and the top comb, form the heart of the combing operation. This is because they are responsible for the fractionation of short fibres from longer and useful fibres. The condition of the wires, the type of half-lap and needles and the needling arrangement, all govern how effectively a comber would do this operation. In addition, the various gauges related to the setting around the cylinder and top comb and the density of spacing of needles also play an important role.

5.1.1 CYLINDER NEEDLING[1]

The normal arrangement of the needles is such that, initially, their action on the lap fringe is gentle and later, gradually intensified. This makes the action of the needles on the fringe more effective. The first few rows (Figure 5.1) of the needles, therefore, gently pierce through the fringe, open it and prepare the fleece for subsequent thorough combing action. This helps in avoiding the suddenness in the combing action, which otherwise is likely to cause fibre damage. If the needles in the beginning rows are closely spaced, they would simply pluck the fibre bunches rather than smoothly going through the fleece and picking only shorter fibres. This invariably leads to fibre breakage and good fibre loss. Equally possible is that the needles simply toss the fringe up, owing to their close spacing, and do not pierce the fleece. Thus, the most ideal arrangement for the needles on the half-lap is that the initial needle rows allow the gradual penetration into the fleece; the subsequent rows go deeper into the fringe and continue the combing action and the last few densely spaced rows, eventually intensify the combing action to complete the job of fractionation. After initial light combing action, therefore, the subsequent needle rows do a more severe and precise job. This is achieved by progressively spacing the needles closer and also making their angle of penetration narrower. This allows each successive needle row to carry out a more thorough job. As regards the height of needles, which decides the overall penetration of the needles into the fringe, it is also progressively increased, but only up to a point. When a certain height is reached, it fairly remains constant thereafter till the final few rows of the needles. The height of these needle rows is only slightly reduced.

It may be now realized that, when the fringe is presented to the first needle row, the fibres in the fleece are comparatively more entangled. A progressive action of the needles with the needling arrangement discussed above, allows each successive row to effectively penetrate the fringe a little more, thus making the whole action smoother. In the absence of such progression, the needles in the first few rows

DOI: 10.1201/9780429486555-5

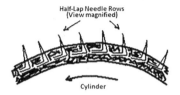

FIGURE 5.1 Cylinder needling:[1,2] The cylinder needles are mounted on a segment called 'half- lap'. On conventional combers, there were as many as 17–20 rows of needles mounted on a half-lap.

would have a tendency to simply bump off the fringe without carrying out the actual combing.

The arrangement of needles on the cylinder, referred to above is analogous to the combing of human hair. The household hair comb that ladies use at home has two distinct divisions – one with coarsely spaced needles and the other with finely spaced ones. It would be easily realized that when the comb is run with a finely pitched tooth portion through the totally uncombed hair, it would exert excessively high forces, thus causing either too much pain for the person (or lady) or resulting in broken hair. The lady, therefore, initially uses the coarsely spaced needle-side of the comb, prepares the long hair and ultimately, turns around the comb to use the finely spaced needles-side to carry out a satisfactory combing job. The same applies to the combing operation too.

As against 20 rows in the conventional comber, the modern comber has only 17 rows of needles. It has been found that, with 20 rows, either the last 2–3 rows do not contribute much in continuing and completing the combing action (they thus become redundant); or if they try to do so, the needles themselves get damaged in turn.

Table 5.1 describes the needling arrangement used in one of the mills in Mumbai for two different systems – Platt's conventional system and the Whitin system.

TABLE 5.1
Needling Arrangement – Count and Gauge[1]

Platt's		Whitin	
No.	Count of needles	No.	Count of Needles
1–4	22	1–3	24
5–7	24	4–7	28
8–9	26	8–10	30
10–11	28	11–17	32
12–13	30	---	---
14–15	31	---	---
16–17	33	---	---
18–20	35	---	---

(*Note:* Increasing the count of the needle means needles are finer)

The ratio of the air space to needle space fairly defines the opening between the needles with respect to needle thickness. The table 5.2 gives an idea of these ratios. As with modern combers, the needle spacing and the needle thickness continuously reduce from the first to the fourteenth row, thereafter, they remain fairly constant.

TABLE 5.2
Ratio of Air Space to Needle Space[6]

Set	Coarser Needles	Average Needles	Finer Needles
1st Row	5.550	4.550	2.380
14th Row	0.430	0.360	0.265

The ratios in Table 5.2 are for a typical comber. It will be clear that the density of the needles, in the beginning, is much less as compared to that in the final few rows (lesser air space). This indicates that the needles in the later rows are spaced much closer to perform more severe and thorough combing action. Whereas the last few rows (Table 5.1) in the conventional comber are quite fine, when the fineness of the corresponding row is compared, the needles in Whitin are a little finer. Again, the fineness of the last few rows is the same in Whitin, however, there is a gradual reduction in the needle fineness in the conventional arrangement. Further, it is found that the needles are thicker and shorter initially. They become finer and a little longer as the rows advance.

With the present trend of processing heavier laps for higher production, the coarsely-pitched arrangement of needles seems to be more beneficial. This is because initially the coarser needles act more effectively in preparing the fringe. At the same time, being sturdier, they are strong enough to bear the strain of heavier fringe.

It is the needle spacing that determines the severity of the combing action. The dimensions of the needles and their projection above the needle bar determine the effective height useful for their penetration into the fringe. When re-needling is done in the mills or when the half-lap is re-needled by the needling company, the specifications as desired for the purpose, are chosen for the effective functioning of the half-lap as a whole. But it is to be understood that there are three variables involved and they interact in this case – (a) number of needles per inch (b) size or the gauge of the needles and (c) projection of needles above the needle strip.

It is, therefore, rather difficult to design a satisfactory half-lap which can meet all possible requirements. Sometimes, some modifications are made especially to overcome the processing difficulties. Equally possible is the fact that some re-needlers substitute the design with small changes when they are out of stock for certain types of needles. They may be even biased for making the changes on the finer side. Usually, they do this assuming that – (a) there is little difference in the dimensions of two types of finer needles and (b) the replacement is in accordance with a general basic theory of using finer needles in subsequent rows. However, incorrect replacement of this kind disturbs the needle spacing, and in all likelihood, may make one or two rows nearly ineffective.

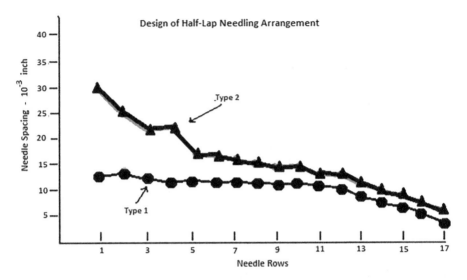

FIGURE 5.2 Needle spacing with different needling arrangements:[2,6] Normally, the initial rows of needles are a little longer, coarser and slightly widely spaced. There is a gradual reduction in the dimensions of needles as the row number increases.

The most accepted criterion is based on the general theory of gradual increase in the intensity of combing action. This means that the needle spacing should gradually decrease from the initial row to the last row. The design of the half-lap needling in type-1 in Figure 5.2 does not seem to follow this fact. Type-2 in this case is comparatively better.

One very useful method in cross-checking the effectiveness of any needling arrangement is to record the whole combing operation with a camera. When such a film is viewed in slow motion, the effectiveness of needle penetration can be easily found. The motion film gives a clear indication of whether the lap fringe allows

TABLE 5.3
A Typical Modern Comber Half-lap Specifications[6]

Strip No.	Needles per Inch	Gauge or Size	Projection Inches	Needle dia. Inches	Needle dia. at Root or Solder Line	Space between Needles at Root
1	26	22	0.156	0.031	0.026	0.0122
2	28	24	0.148	0.024	0.020	0.0154.
3.	30	24	0.140	0.024	0.020	0.0133
7.	38	24	0.124	0.015	0.014	0.0123
10.	48	29	0.112	0.014	0.011	0.0090
13.	64	31	0.100	0.012	0.010	0.0052
16.	82	32	0.084	0.011	0.009	0.0031
17.	88	32	0.078	0.011	0.008	0.0026

satisfactory penetration. It is here that the design of initial rows is important. If the needle spacing in the beginning is comparatively narrow, it simply tosses the fringe. Sometimes, the change in the angle of the first few rows is tried as one of the solutions; but it involves altogether a different design of the needles. The fringe not allowing the needles to gently penetrate may be possibly also due to the unfavourable needle spacing.

On the basis of such observations, it is possible to redesign the whole needling arrangement. Yet another needling design (Type-3, Figure 5.3) was tried on one of the combers. The new half-lap when put on this comber, not only improved the combing performance but also the 'bouncing' effect almost disappeared. The revision thus resulted in a significant reduction of comber noil. A proper selection of half-lap needling is, therefore, a very important criterion on which the combing quality depends. If the mill is more than satisfied with their quality level, the saving can then be made by reducing comber noil. If the mill decides to replace their existing half-laps with new different specification, the change can be brought over as and when the old half-laps become unserviceable. The saving in the waste that would be realized then will be made at no extra cost. However, it must be emphasized that the amount of saving in the comber waste by making suitable changes in the half-lap needling solely depends upon the mill's existing level of waste extraction, the cost of cotton and the present efficiency level of existing half-laps.

The modifications in the needling arrangement must also be viewed from the point of view of needle damage. The changeover from type-1 to type-2 (Figure 5.2) may give reduce damage to the needles as it allows free and smooth passage for the needle rows to pass through the lap fringe. In a way, therefore, it may also reduce the expenditure on needle maintenance. This is exactly what has to be done when thinking of processing heavier laps. Especially on high-speed combers where the cylinder needle action is going to be much faster, any stress on the needles, due to heavy laps

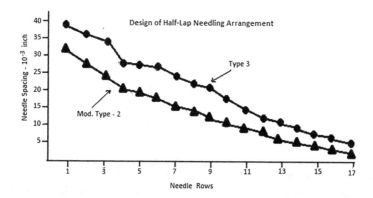

FIGURE 5.3 Needle spacing with different needling arrangements:[6] There is always a difference of opinions as to which needling arrangement is best. Apart from the basic function of removing short fibres, a typical arrangement must protect the fringe from fibre damage.

can prove to be detrimental. In such cases, smooth, gentle and yet effective penetration of needles is going to be much more beneficial.

There is another angle to this whole issue. When carrying out the modifications in the cylinder needling arrangement, it was thought that rather than trying to modify the needling arrangement, it would be really worth trying to make the construction of the whole half-lap much stronger to sustain the stress. This was necessary, especially at higher speeds and that too with heavier comber laps. In this respect, the introduction of Uni-Comb found its way through and was universally accepted for processing heavier feed.

5.1.2 Top Comb Needling[1,2]

The top comb needle spacing is much wider as compared to cylinder needling. As many as 10–15 fibres, can pass through the top comb needle space. Under no circumstances, the needling arrangement is capable of individualizing the fibres. It is, therefore, expected that slight differences in needle spacing should not critically affect the amount of waste removed; although very wide spacing of the needles may reduce waste significantly. An experiment carried out in this respect led to the results in Table 5.4.

It is, however, observed that the effect of this spacing of the needles is better seen at a wider step gauge where closer needle spacing gives higher waste extraction as compared to wider spacing. The closer spacing also results in better short-fibre removal owing to their better restraining power over the fibres in general. With closer spacing, it has been reported that the mean length improves and there is better nep reduction.

TABLE 5.4
Top Comb Needle Spacing & Waste

Needles/inch	Spacing in 1/1000″	Waste%
45	7.5	12.5
56	6.5	12.5
66	5.0	13.6

5.1.3 Uni-Comb[2,3]

Both, the type and the construction of the half-lap in the Uni-Comb (Figure 5.4), are very much different from that in the conventional half-lap. The rows of needles are totally dispensed with and are replaced by a segment holding metallic saw-tooth wires. The metallic wires on a segment are arranged in a slanting position so that nothing across the width of the Uni-Comb is missed. This helps in improving its performance.

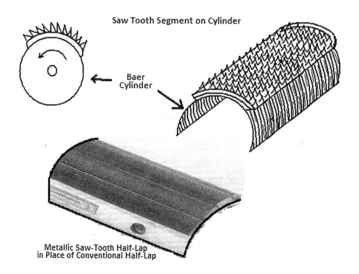

FIGURE 5.4 Uni-Comb half-lap:[2,6] As against the fine needles in conventional half-lap, the Uni-Comb has a fine saw-tooth wire segment. While comparing their performance, it is important to note their wire point density.

This segment, like a conventional half-lap, can be mounted on the bare cylinder. However, unlike conventional needles, the metallic saw-tooth wires do not retain fibres on their surface. Therefore, when they are used, the comber noil is reduced.

This is because the saw-tooth wires barely hold the fibres and it becomes easy for the brush to strip and clean them. This reduces the comber noil. Further, the complete cleaning of wires thus helps in improved combing action in every cycle. It has been found that even with the reduced noil levels with Uni-Comb, its quality performance closely resembles that of a conventional half-lap.

As mentioned earlier, the saw teeth on Uni-Comb are disposed of with a slight angle and certain regularity to provide a very fine combing action. However, the lateral spacing is comparatively wider. This wider spacing together with teeth arranged at an angle offers a smooth combing. It also prevents fibre breakage or damage. The shape of the saw tooth is uniquely designed so that the short-fibre waste does not penetrate deeper into the teeth. This makes the stripping action of the brush more effective. The brush bristles also need only a little penetration to clean the saw-tooth wire segment. This increases their life.

The Uni-Comb, as claimed by the manufacturer, has opened a great possibility of raising the machine's speed. Especially with heavier laps, the increased speed tends to accumulate the fibres on traditional cylinders; whereas, with Uni-Comb this possibility is almost eliminated. So also, the saw-tooth wires are strong and sturdy, they hardly experience any strain even with heavier laps. However, it is important that when finer cottons are required to be combed, it is necessary to use a finer set of saw-teeth wire segments with higher point density. In the latest versions of Uni-Comb, the segment is divided into two to three compartments with varying density and fineness of saw teeth. This has considerably improved the performance of Uni-Comb. With conventional half-laps however, the lap weight and the speed of the machine when processing finer cottons were required to be reduced. The introduction of Uni-Comb removes such restrictions owing to its solid and sturdy construction.

5.1.3.1 Performance of Uni-Comb[4,5]

The relative performance of Uni-Comb, like conventional half-laps, depends upon such factors as the condition of the wires, their specifications, point density, etc. Some local manufacturers in India, have introduced 'metallic comb'. It is claimed that these combs have rich pin density and finer pinning patterns, thus improving their combing action. It is further claimed that such a finer arrangement leads to improved CSP and reduction in neps level in the final yarn.

In the past, research trials were conducted, comparing the performance of Uni-Comb with conventional half-lap. Different combing parameters, such as waste%, lap weight, feed length, etc., were varied. Even different mixings were used to judge whether and where Uni-Comb would be most suitable for a certain range of mixings.

It was found[3] that the performance of Uni-Comb, in terms of neps, yarn evenness and yarn strength, was comparable to that of conventional half-lap. However, it was observed that Uni-Comb was found to work better with medium mixings. However. when finer cottons such as Giza-45 were used, the strength, evenness and appearance of yarn were found to deteriorate.[4] It is now known that some of these difficulties have been overcome by the manufacturers by providing new designs of Uni-Comb. With conventional half-laps, longer feed length improved the mean length of the combed mixing. However, longer feed length is not found to be beneficial with Uni-Comb. With higher lap weights, the nep removal efficiency of conventional half-laps was found to deteriorate; whereas saw teeth in Uni-Comb being strong and sturdy, the rise in the nep level was not significant. A typical bouncing effect, however, was noticed in some instances with Uni-Comb. As against this, the typical arrangement of half-lap needles, their spacing, etc., enabled a comparatively smooth penetration of needles, reducing a possible tendency for 'bouncing-off'.

The yarn evenness and imperfections in the yarns spun from the combed material produced with both Uni-Comb and conventional half-lap were comparable. The studies on Yarn Quality Factor (YQF) showed that Uni-comb performed well with both lighter and heavier lap feed.

$$YQF = \frac{C.S.P. \times 100}{Total\ Imperfections \times U\%}$$

As mentioned earlier, the Uni-Comb segment, such as licker-in wires, is composed of hard saw-teeth wire segments, where the wire points on adjacent wire on the segment are staggered and arranged in twill lines running at 45° to the combing direction. Also due to the fact that the points per unit area are the same throughout, the action of Uni-Comb is bound to be sudden and severe right from the beginning.

The fibres are thus abruptly brought under the action of a very high number of wire points per unit area. Further, the number of wire points per unit area is yet another criterion which seems to be important when processing different mixings varying in grades. For example, with finer mixing, the number of fibres in the lap of a certain weight (g/m) would be much higher, demanding more wire points for effective combing action.

Some experiments were carried out[5] on these lines where two different types of Uni-Combs were used such that one of them (1500F) had a total number of 8,528 points on the whole segment; whereas the other (1800) had a total number of 14,703 points. As against this, the conventional half-lap had a total number of 11,519 points on the whole segment. The point density for each of these segments is as follows:

(1) No. 1500F – 254 points per sq. inch; (2) No. 1800 – 384 points per sq. inch

The conventional half-lap had 100 points per sq. inch at the start and 380 points per sq. inch at the end.

It was found that in the case of certain yarn properties like evenness, strength and imperfections, the performance of Uni-Comb No. 1800 was not only comparable, but it even excelled. It was then thought that this Uni-Comb No. 1800 providing a sufficiently high number of wire points throughout the cycle, as compared to those provided by conventional half-lap, might be the reason for its excellence. As compared to No. 1800, however, the performance of No. 1500F was not satisfactory.

It can therefore be concluded that, like conventional half-lap needling, it is equally important to select Uni-Comb with proper wire specifications, especially the number of wire points per sq. inch. Thus for finer mixing, requiring more point density, an appropriate Uni-Comb needs to be chosen to maintain its performance.

REFERENCES

1. *Manual of Cotton Spinning: "Draw Frames, Comber & Speed Frames": Frank Charnley*, The Textile Institute Manchester, Butterworths, 1964
2. *Elements of Cotton Spinning – Combing - Dr. A.R.Khare*, Sai Publication
3. Uni-Comb – Nitto Shoji Ltd. Booklet
4. S. Y. Nanal – CQC report, Mafatlal Group 21/1169
5. A. R. Khare – Indian Textile Journal, July 1977, p.83 & Nov. 1980, p.127
6. Effect of half-lap needling on combing efficiency: P.E.Sperling, T.R.J. 1958

6 Other Aspects

6.1 OTHER ASPECTS

6.1.1 DOUBLE COMBING

It is a universally accepted fact that combed yarns are better in their appearance than carded yarns. Normally, the material is combed only once. But, if the removal of short fibres is the only criterion, perhaps a single combing operation would suffice the purpose adequately. Some yarns, however, demand a much better appearance wherein, apart from yarn uniformity, the overall yarn grade is required to be improved (in terms of the neps and hairiness). The idea of double combing was conceived to meet these demands. Presently, double combing, or sometimes referred to as 'super combing', is followed for very high grade and superfine products such as voiles and mulls. This is because the yarn and, therefore, the fabric appearance in such varieties are of utmost importance. Technologically also, it has been proved that waste extraction beyond 18%, mostly goes in to improve the appearance of yarn and therefore that of fabric.

Generally, double combing becomes necessary when the waste extraction levels are above 20%. It may also be mentioned here that waste levels above 18%, though possibly can be attained in one single operation, will not be economical from the point of view of the loss of good fibres during the operation.

The normal process for the second combing operation is to make again a lap using suitable preparatory machinery. This preparation is not like the one discussed earlier in this book. There is no demand for draft or parallelization or even direction of the feed. Even then, there are other difficulties. As the material for making the laps for second combing is the soft and smooth comber sliver, the major problem commonly faced is the handling of such delicate and weak sliver. This needs maximum care in handling the material – right from feeding the combed sliver to suitably making the laps on a lap-making machine. With a very weak sliver, even the smallest possible stretching during lap-making leads to an unevenness of sizable magnitude. This has to be avoided.

The stretching may be in the form of a false draft on the feed table or withdrawal of slivers from feed cans. The can springs are required to be in top condition. Even the creel height has to be reduced for this purpose to avoid any stretching due to hanging sliver weight. The false draft on the table is usually due to a rough tabletop. The table surface needs to be highly polished. For arriving at the desired lap weight, the hank of sliver at first combing must be accordingly arranged. The sliver after the first combing is highly parallelized. The draft, in making a lap for the second combing, therefore, needs to be kept at a minimum. In fact, no further parallelisation is

DOI: 10.1201/9780429486555-6

required. Even then, the minimum level of the draft does help in merging the slivers with their neighbours. This is important for making a good lap. Maintaining a minimum draft level also helps in reducing any lap-licking tendency. For a total waste of 25%, the noil extracted in the first combing is around 17–18%, and that in the second combing is around 7–8%.

Double combing improves the cotton fibre characteristics in respect of mean length. The short-fibre content is substantially reduced as compared to single combing. Another important and marked improvement is that there is a significant reduction in neps. A little improvement is also seen in lea CSP and single-thread tenacity. However, the yarn elongation is not much changed. The fineness in terms of micronaire of combed cotton is, however, not affected by double combing.

In short, the double-combed yarns excel in comparison to those made from single combing, simply due to their superior appearance arising out of improved evenness and substantial nep reduction.

6.1.2 COMBING OPERATION AND QUALITY[3,11,12]

There is a general improvement in the yarn quality when cotton is combed. Even with 10% comber noil extracted, the yarn strength ((3) in Figure 6.1) is approximately 10% higher than the corresponding carded yarn. However, the increase in the yarn strength beyond 10% noil, does not seem to be much.

Improvement in the yarn evenness ((2) in Figure 6.2) is comparatively better. The improvement in the yarn imperfection with an increase in the noil is comparatively far better. Till about 15% noil, the improvement is rapid. Though the improvement between 15% and 20% is comparatively less; still it is noticeable. Thus, the yarn quality improvement is not linearly related to comber noil. The ring frame performance in terms of ends-down is generally improved with combed yarn. Thus, the end breakage rate is lower with combed yarn. This reduction is also not linearly related.

6.1.3 PROBLEMS IN COMBING BLENDS[5]

The combing of blends consisting of a mixture of heterogeneous fibres is not common in the textile industry. The entire process of combing sorts out the fibres of different lengths in such a fashion that the fibres greater than the detachment setting (minus feed length) go into making a comber sliver (in the forward-feeding method). This sorting also depends upon the Baer sorter diagram of the material being processed. It is evident from the diagram (Figure 6.1) that the length of the cotton fibres is not uniform all throughout. The typical nature of the Baer sorter diagram for cotton owes its origin mainly to the fibre growth on cotton seeds in the field. For viscose fibro, however, the manufactured length is the same for all the fibres. Even then, there is some fibre breakage that occurs mainly in the earlier processes (blow room and card). Thus, there is a small percentage of fibres shorter than the cut staple length.

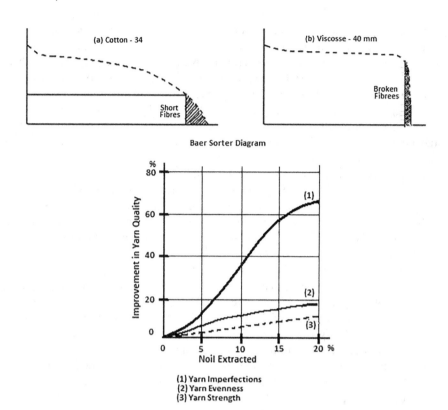

Baer Sorter Diagram

(1) Yarn Imperfections
(2) Yarn Evenness
(3) Yarn Strength

FIGURE 6.1   (A) Baer sorter diagram of the two components:[2,13] As against the tapering curve with cotton, indicating some proportion of short fibres, the curve for manmade fibre is flat, showing uniformity in the length. (B) Effect of comber noil% on yarn quality:[2,11] In comparison to significant improvement in yarn imperfections, when the comber noil is increased, there is less improvement in yarn evenness and strength.

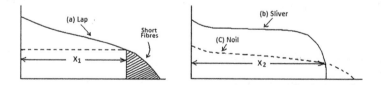

FIGURE 6.2   Baer sorter diagram – before and after combing:[11,13] This is one of the methods of judging the comber performance. With the removal of short fibres, the ideal Baer sorter diagram should have shortened tail.

Therefore, there will be some proportion of man-made fibres in the noil. But the proportion is expected to be much smaller as compared to that for the cotton component with which it is blended. However, when blended laps were fed to the comber, the noil contained some sizable proportion of viscose fibro component. This confirms again that one of the reasons for long-fibre loss in combing is due to the premature dragging of fibres ahead during detachment. It is also possible that the process of combing itself involves some fibre breakage. The loss of a valuable proportion of staple fibres (fibro), in this manner, very much restricts the use of the combing process for processing blends.

In fact, apart from blow room blending, the comber could have been another ideal place to mix the heterogeneous fibres. It is also apparent that when cotton is blended with any known manmade staple fibre, the number of faults would substantially reduce to give a better-combed web. Even the use of a wider step gauge, to remove more short fibres from cotton, would not pose any problem for a manmade component.

However, it is found that when the step gauge is widened, the proportion of staple components in the noil also increases. Further, it is difficult to maintain the desired proportion of components of blends in the combed sliver as compared to what is fed. This is because the extraction of the short fibres is mainly from shorter components (cotton). In a 67: 33 polyester/cotton blend ratio, the noil extracted contained much less proportion of polyester, thus resulting in higher polyester content in the combed web.

One of the reasons assigned to fibre breakage of manmade components in combing is that a disproportionately smaller feed length is required to be used for shorter-length cotton components. Even the greater bulk of manmade fibres in comparison to cotton, poses problems during their movement through the cylinder and top comb needles. This again leads to fibre breakage. Further, owing to a difference in bulk, the nippers may not be able to grip both the component fibres satisfactorily.

Even then there are certain interesting possibilities when one thinks of combing a blended material. The appearance of manmade fibres in comber noil, especially if they are thermoplastic, can facilitate their use in non-wovens. Thus, the noil could be used for producing non-wovens with thermo-bonding or with needle punching if the manmade fibres are non-thermoplastic. Further, the comber noil containing a small percentage of manmade components could be used for producing special yarns, thus fetching either more price or adding to some fancy effects. Another possibility is to use noil to make open-end yarns.

6.1.4 FRACTIONATING EFFICIENCY

The process of combing involves the loss of material. It is, therefore, necessary to arrive at a condition that gives a minimum loss of good fibres while achieving maximum short-fibre removal to get substantial upgrading of raw material. The economics of the process of carrying out this largely depends on its efficiency.

In the industry, the intensity of combing is expressed in terms of the percentage of noil removed. The common terms used are – 'scratch combing', 'semi-combing', 'normal combing' and 'super (double) combing'. The most common and easiest method to judge the up-gradation achieved is to define the change in the fibre length parameters – mean length and short fibre%, before and after the process. Thus, a minimum of 2 mm improvement in mean length could be considered satisfactory for a combing performance. In addition, the Baer sorter diagram for the lap, sliver and noil also can depict the nature of fibre length and its distribution. It can be observed (Figure 6.2) that, the predominant taper at the end of the curve in the case of lap is absent with a combed sliver.

Similarly, the difference in the curves for sliver and noil would also throw some light on the efficiency of the process. Another criterion for judging the combing performance is to find the nep reduction through the process. Even the yarn appearance could be observed for improvement in yarn grade. All these methods are sensitive and related to noil level and machine parameters. They, however, do not reflect upon the performance of a comber in terms of how many short fibres have been passed on to the comber sliver, or to that extent, how many long fibres are lost in the comber noil. Therefore, they are directly not useful in studying the actual combing efficiency in terms of actual short-fibre removal or the short fibres passed on to comber sliver.

6.1.4.1 Different Methods of Finding Fractionating Efficiency

The concept used in all these methods is how best a comber can remove short fibres less than a certain length, usually termed as 'boundary length'. This length is arbitrary, in the sense that, it is assumed depending upon the effective length of the fibre. Thus, with a higher effective length, the boundary length is also higher. There is another angle to the functioning of the comber. As the comber is expected to remove fibres shorter than a certain length, it is also expected to safeguard all those fibres which are long enough. This again is conceptually based on the boundary length. It can therefore, can be understood that both these functions are equally important for the efficient functioning of a comber.

6.1.4.2 Simpson and Ruppennicker Method[8]

In this method, the distribution of the percentage of fibres 'a' corresponding to a mean group length (ℓ) for comber laps was found. Similar values for comber sliver [A = % fibres for mean group length L] were found. In this way, we would have –

a_1, a_2, – am, – ax [as percentage of fibres for lap] and
A_1, A_2, – Am,– Ax [as percentage for sliver]

Note: A_1, A_2, etc., are the corresponding percentage of the original lap distribution. Similarly,

ℓ_1, ℓ_2,– ℓm,– ℓx [corresponding mean group length for lap] and
L_1, L_2, – Lm,– Lx [corresponding mean group length for sliver]

Note: For all practical purposes (ℓ_1, ℓ_2, ...) and (L_1, $L_{2,...}$) are the same. Also, ℓm and Lm as the boundary lengths for lap and sliver are also the same. It is expected that a comber removes all the fibres shorter than this boundary length as a noil.

As one can expect, the percentage of short fibres in the above fibre distribution to form comber noil would be as follows:

$$(a_1 - A_1), (a_2 - A_2), (am - Am), (ax - Ax).$$

For ideal noil, however, the fibres below the boundary length should only be removed. But the observations reveal that there is a small percentage of fibres longer than boundary length in the noil.

Note: As, (a₁ to am) represents the percentage of length groups up to boundary length in the lap and (A₁ to Am) represents the percentage of length groups in sliver (expressed as a percentage of original lap distribution),

$\sum [(a - A) / \ell]$ gives the short-fibre distribution in actual noil and $\sum (a / \ell)$ gives the distribution for ideal noil. The combing efficiency is then defined as:

$$\text{Combing Efficiency} = \frac{\% \text{ of fibres shorter than } \ell m \text{ in } \mathbf{actual} \text{ noil}}{\% \text{ of fibres shorter than } \ell m \text{ in } \mathbf{ideal} \text{ noil}}$$

$$\text{Combing Efficiency} = \frac{\sum_{\ell_1}^{\ell_m} \frac{a - A}{\ell}}{\sum_{\ell_1} \frac{a}{\ell}} \times 100 \tag{1}$$

Here the summation is taken over the range ℓ_1 to ℓm as we are only concerned with the percentage of short fibres shorter than the boundary length. Also, the division by 'ℓ' gives the due weightage to each length group, in particular to the shorter fibres.

Similarly:

$$\text{Detaching Efficiency} = \frac{\% \text{ of fibres longer than } \ell m \text{ in the } \mathbf{sliver}}{\% \text{ of fibres longer than } \ell m \text{ in } \mathbf{lap}} \times 100$$

$$\text{Detaching Efficiency} = \frac{\sum_{\ell_m}^{\ell_{max}} A \times \ell}{\sum_{\ell_m}^{\ell_{max}} a \times \ell} \times 100 \tag{2}$$

Note: Here too, the summation is taken over the range ℓm to ℓmax as we are only concerned with the percentage of fibres longer than the boundary length. Also, the multiplication by 'ℓ' gives the due weightage to each length group, in particular to the longer fibres.

It is often criticized that these two ratios – 'combing efficiency' and 'detaching efficiency' are two different indices and not related to each other. This is because two different length groups are involved in the computation. It may thus be possible that when two different combers are compared for their performance, the values obtained for either detaching efficiency or combing efficiency level may be identical but the values for the other index could materially be different, thus leading to ambiguity. Detaching efficiency can vary from (100 − waste%) to 100%; while combing efficiency varies from (waste% to 100%).

6.1.4.3 Parthsarathy Method[7]

The composite index, taking into account both the percentage of long fibres retained in the sliver and short fibres entering the same sliver are a better measure according to the author. This method is based again on a similar procedure wherein the overall distributions of all the fibre length groups in the comber sliver and that in the lap are found. In this method, the fractionating efficiency index (FEI) is composed of two factors – one due to efficient detachment of longer fibres (a positive contribution from comber) and the other due to inadvertent withdrawal of short fibres into comber sliver (a negative contribution, reducing (FEI)). The various parameters used are as follows:

a_1, a_2, − am, − ax [as a percentage of fibres for lap] &
ℓ_1, ℓ_2, − ℓm, − ℓx [corresponding mean group length for the lap]
A_1, A_2, − Am, − Ax [as percentage of fibres for sliver]
L_1, L_2, − Lm, − Lx [corresponding mean group length for the sliver]

ℓm and Lm – are the boundary length below which the fibres should go into waste and W is the noil percentage extracted.

Note: The values of A_1, A_2, − Am, − Ax are calculated from the actual values − A'_1, A'_2, etc. of the comber sliver which are multiplied by a factor (100 − W).

F.E.I. for Longer Fibre Retention in Combed Sliver:
$\sum$ (Lx − Lm).Ax − (1) Contribution of fibres longer than Lm in the **Combed Sliver**
In the same way:
$\sum$ (ℓx − ℓm).ax − (2) Contribution of fibres longer than ℓm in the **Comber Lap**
The limits over which the summation is carried out are from ℓm to ℓmax or Lm to Lmax

Thus, **FEI for Longer Fibre** retentions will be as follows:

$$F(i) = \frac{\sum_{\ell_m}^{\ell_{max}}\left(\ell_x - \ell_m\right)Ax}{\sum_{\ell_m}^{\ell_{max}}\left(\ell_x - \ell_m\right)ax}$$

Note: The length groups are the same for both the lap and the sliver. Also, Ax stands for percentage of fibres for sliver, whereas ax stands for the percentage of fibres for the lap.

F.E.I. for Short Fibres in Combed Sliver:
$\sum$ (Lm − L_1).Ax − (4) Contribution of fibres shorter than Lm in the **Combed Sliver**
Again, in the same way,
$\sum$ (ℓm − ℓ_1).ax − (5) Contribution of fibres shorter than ℓm in the **Comber Lap**

The limits over which the summation is carried out are from ℓ_1–ℓm or L_1–Lm
Thus, **FEI for Short Fibres** in Combed Sliver will be as follows:

$$F(ii) = \frac{\sum_{\ell_1}^{\ell_m}\left(\ell_m - \ell_x\right)Ax}{\sum_{\ell_1}^{\ell_m}\left(\ell_m - \ell_x\right)ax}$$

Therefore composite FEI will be = [FEI.(i) – FEI.(ii)] × 100

In an ideal working, FEI.(i) is 1; while FEI.(ii) is 0, thus a composite FEI will be 100 in an ideal condition.

6.1.4.4   Owalekar Method[6]

Three concepts were introduced in this method: (1) preferential short-fibre removal, (2) preferential long-fibre retention and (3) waste removal efficiency.

The short fibres are those fibres which the machine is supposed to remove with the given parameters and under ideal conditions. The short-fibre removal, therefore, is defined as the ratio of the difference of short fibre% in the lap and sliver to the short fibre% in the lap, expressed as a percentage. Similarly, the long-fibre retention is a ratio of long fibres% in sliver and that in the lap (expressed as a percentage).

There are two terms – 'actual noil' which the comber actually removes and 'theoretical noil' which is based on the short-fibre content in the lap fed to the comber. The waste removal efficiency is the ratio of actual noil% to the theoretical noil%, expressed as a percentage.

It is really important to know how best a comber 'preferentially' removes the short fibres with a certain precision. Similarly, it will be equally important to see how a comber incurs minimum good fibre loss and very 'effectively' retains the long fibres.

Logic – fibres going into noil or forming sliver: In the backward feed, let it be assumed that the feed roller feeds the length 'f', when the nippers recede back. If 'x' is the fraction of this delivery taking place before the nippers close for combing by cylinder, then 'fx' is the length of the fringe which would additionally project in front of the nipper grip-line for the following combing cycle. Also, if the detachment setting is 'd', the projected length of the fringe between the nipper and detaching roller at the completion of detachment would be 'd'. Therefore, the total length of the fringe projecting in front of the nipper grip-line at the time of cylinder combing would be (d + fx).

This means that, during the cylinder combing operation, all the fibres longer than (d + fx) would be held firmly by the nipper grip and hence in subsequent piecing operations, would go into the sliver. Thus, the projecting length after combing would be also (d + fx). During detachment, however, the remaining (1 – x) fraction-length of feed (i.e., (1 – x).f) which remained behind the nippers and got accumulated after the nipper closure for cylinder combing, is now available. Owing to the advancement of this same length during the forward movement of the nippers for the piecing-up operation, the apparent distance between the nippers and detaching line will be reduced by this margin. Hence, this apparent distance at the time of detachment will be:

$$[d - (1 - x) \cdot f] \text{ OR } (d + fx - f)$$

This means that fibres shorter than $(d + fx - f)$ would go into noil. This is because they will not be allowed to reach the detaching line owing to the presence of the top comb. Thus:

Fibres > $(d + fx)$ – not removed during cylinder combing, would go into sliver
   and
Fibres < $(d + fx - f)$ – would go into noil.

The fibres of length in between $(d + fx - f)$ and $(d + fx)$ would either go into sliver or noil, depending upon their disposition and the influence of neighbouring fibres. So it can be safely assumed that a boundary at $[d + fx - (f / 2)]$ length divides the shorter and longer fibres, going into noil and sliver respectively. The error involved in the assumption, such as this, is negligible. It is also required to assume that there is no fibre breakage occurring in the comber.

The following equations are put forth by Owalekar[6]:
Let it be assumed that N = % of actual noil, then
100 g of lap cotton when fed to comber, will deliver – $(100 - N)$ g of sliver – (1)
If Ts = % of trash and neps in comber sliver
   100 g combed sliver will have $(100 - Ts)$ g of actual fibres in the sliver – (2)

Hence, in $(100 - N_1)$ g of combed sliver will have, $\dfrac{(100 - N)(100 - Ts)}{100}$ g of actual fibres in sliver form – (A)
Now, if T = % of trash and neps in comber lap
   The actual fibre content in the lap cotton will be – $(100 - T)$ g – (3)
This means that:
$(100 - T)$ g of actual fibres in the lap gives – [$(100 - N)(100 - Ts)$] / 100 g of fibres in combed sliver – (B)
The actual noil (N%), therefore, is required to be corrected.

$$\textbf{Corrected Noil } \% \left(N'\right) = \left\{ (100 - T) - \frac{\left[(100 - N)(100 - Ts)\right]/100}{(100 - T)} \times 100 \right\} --(1)$$

When actual noil (N%) is corrected as N′, the Waste Removal Efficiency (WRE) can be defined as:

   WRE = [Corrected Noil / Theoretical Noil] $\times$ 100 = (N′ / N) $\times$ 100 – (2)

Note: The theoretical noil N is the actual short fibre% in the lap cotton. It is those fibres which are less than a boundary length 'ℓ' which is equal to [$d + fx - f / 2$] as mentioned earlier.

   Further, If we have:
   Ws = Actual% of fibres shorter than 'ℓ' as from Baer sorter diagram of comber sliver, then:

When 100 g of sliver gives Ws g of fibres in the sliver shorter than 'ℓ', then:

(100 – N) g of sliver would give {[(100 – N1).Ws / 100] g of fibres shorter than 'ℓ' in the sliver. The short-fibre removal efficiency, therefore, will be:

$$\textbf{Short Fibre Removal Eff.} = \frac{N - \left[\left(100 - N_1\right) \times Ws\right]/100}{N} \times 100 - - - (3)$$

Note: N = % of short fibres as from Baer sorter diagram of lap

With Ws % of fibres shorter than 'ℓ' in the sliver, (100 – Ws) % of fibres will be longer than 'ℓ'. Also, for every 100 g of lap cotton, there will be (100 – N) g of fibres in the sliver. Thus, for every 100 g of lap cotton, there will be:

[(100 – N₁) (100 – Ws)] / 100 g of longer fibres in the sliver.

Theoretically, with N% of fibres shorter than 'ℓ' in the comber lap (which is short fibre% from Baer sorter diagram), there will be (100 – N) g of longer fibres in the lap.

The long-fibre retention efficiency will be:

$$\textbf{Long Fibre Retention Eff.} = \frac{\left(100 - N\right)\left(100 - Ws\right)}{100 \times \left(100 - N\right)} \times 100$$

$$LFR = \frac{\left(100 - N\right)\left(100 - Ws\right)}{\left(100 - N\right)} - - - (4)$$

Preferential Short-Fibre Removal Efficiency (P.S.F.R.E.):

$$\frac{S.F.R.\% - N}{\left(100 - N\right)} \times 100 - - - (5)$$

This gives the level of short fibres removed preferentially. If S.F.R. = N there is no preferential short-fibre removal. If S.F.R. = 100, then P.S.F.R.E. = 1, which indicates complete removal of short-fibre removal.

Preferential Long-Fibre Retention Efficiency (P.L.F.R.E.):

P.L.F.R.E. gives the efficiency of the machine in saving long fibres. If, L.F.R. = (100 – N), there is no preferential long-fibre retention and P.L.F.R.E. = 0

However, if L.F.R. = 100, then P.L.F.R.E. =1 which means that the process does not involve the loss of long fibres.

$$P.L.F.R.E = \frac{L.F.R.\% - 100 + N}{N_1} \times 100 - - - (6)$$

6.1.4.5   Importance of the Ratios

As reasoned out by the author, there are several reasons for lower values of fractionating efficiency. The removal of short fibres basically depends upon the action of combing by the cylinder. Here the condition of the half-lap needles and their inadequate penetration are the main reasons for leading to poor fractionating efficiency. Deficient lap preparation leads to high disorientation of fibres. The higher entanglement of fibres in the lap allows harsher action of half-lap or top comb. Even the presence of hooked fibres plays a significant role. Ineffective nipper grip, wrong settings or even bad conditions of the top detaching roller covers can seriously affect fractionation.

Loss of long fibres mainly occurs due to inadequate grip of the nippers. A bad condition of the top comb or unsatisfactory lap preparation invariably leads to the breakage of long fibres in combing. The loss of long fibres in the comber waste can be both seen and felt. Their presence in the comber noil changes the appearance of noil. So also, the noil will have a softer feel. It should also be borne in mind that the two values – 'short-fibre removal' and 'long-fibre retention' – will be related only when there is **no** fibre breakage. Even then, P.S.F.R.E and P.L.F.R.E. are independent of noil level and can be used to compare combing action at different noil levels and settings.

6.1.4.6 Factors Affecting Fractionating Efficiency[11,13]

6.1.4.6.1 Comber Waste Extraction

It is known that increasing the step gauge increases the comber noil. This affects the fractionating power of the comber. However, this increase in power is linear up to approximately 20% noil. At higher levels, the fractionating power of the comber seems to deviate because of the nature of fibre length distribution. In general, preferential long-fibre retention (P.L.F.R.E.) improves at wider step gauge.

However, with a very wide step gauge, P.S.F.R.E. deteriorates. This is especially true when working with the top comb. Nep removal efficiency in presence of a top comb improves with a higher step gauge.

6.1.4.6.2 Type of Feed

It has been already mentioned (Section 4.6.2.2) that the waste extracted in backward feed is more. However, it has been observed that the fractionating efficiency is better with forward feed. Even when the waste levels and feed lengths are kept identical, still fractionation is better with forward feed. In general, the longer feed length decreases noil% and gives poor P.S.F.R.E. There is no effect of change in the feed length on nep removal efficiency.

6.1.4.6.3 Mean Length

Fractionation increases linearly with improvement in mean length. At a wider step gauge, the improvement in mean length is better. The absence of the top comb has generally led to a lower mean length.

6.1.4.6.4 Lap Preparation

It has been observed that it is possible to improve the fractionating efficiency index by using a better lap preparation system. However, with heavier lap weights, preferential short-fibre removal efficiency is usually poor.

6.1.4.6.5 Top Comb

The effect of top comb on fractionation is typically complicated. When it is set too deep into the fringe, it gives poor fractionation. This is because, in this case, it leads to fibre rupture. When the top comb is removed, it also leads to poor fractionation. This is because the waste% is drastically reduced. Without a top comb, the fibres, especially the shorter ones, get accelerated and pass on to form the sliver. Obviously, in such a case, the preferential short-fibre removal efficiency is poor.

At a higher production rate, there is deterioration in fractionation, even when the top comb needle specification is maintained the same. This is owing to long-fibre breakage during detachment.

6.1.5 Combing and Cutting Ratio[9]:

The card sliver has the most random arrangement of fibres. As discussed in the earlier chapter on lap preparation, it is very important, to give attention to the improvement in fibre orientation through pre-comb drafting. Also, the number of hooks, their extent and the direction of the presentation are equally important. In this connection, a method was devised by Lindsley[9] to measure the 'fibre orientation'. This method has been very useful in finding the improvement in fibre orientation (whether random) and fibre parallelization. The two ratios – combing ratio and cutting ratio (orientation index) – give an idea of the estimate of the randomness of fibre orientation and the proportion of hooks and their extent in leading or trailing direction. The ratios are thus useful in judging the performance of pre-comb processes in terms of an improvement in fibre orientation (or reduction in both the number and extent of fibre hooks).

6.1.5.1 Description of the Apparatus

Base Block P and Clamping Plate XY (Figure 6.3) are used to mount the material (sliver or roving). Cutting Plate ABCD is mounted with its faces CD perfectly machined to align, face to face, with the edge of Clamping Plate XY. The plates XY

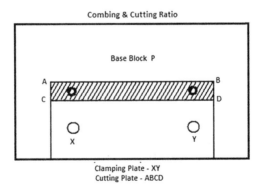

FIGURE 6.3 Lindsley's apparatus: This had been developed basically to find the fibre orientation in any given material. One of the findings related to fibre orientation in card sliver was that the fibres were most randomly orientated.

and ABCD are mounted on Base Block P. Both these plates are provided with holes at their ends to fit over the pins carried by the base block.

They serve to align the two plates perfectly and also brings them back to the same initial position when they are removed during the experimentation. The lower side of Clamping Plate XY is recessed slightly so as to provide a better hold on the sample when mounted over the base block.

Cardboard is laid on the base block. In the normal non-working position, the plates XY and ABCD rest on this cardboard. When a sample (sliver or roving) is required to be examined, both Cutting Plate ABCD and Clamping Plate XY are first removed. The sliver or roving sample is then laid over the cardboard and both plates are then replaced back in their respective positions.

This leaves the edge of the sample exposed as shown in Figure 6.4. With a sharp razor blade, the sample is trimmed at the edge AB of the cutting plate. The trimmed portion is discarded.

Clamping Plate ABCD is removed and the exposed fibre fringe under it is gently, gradually and carefully combed-out. It thus straightens and parallelizes the fibres (Figure 6.5). The combed-out portion (the one carried by the brush bristles) is collected and weighed (C).

After this, Cutting Plate ABCD is again replaced back and the fringe extending beyond AB (Figure 6.6) is trimmed and weighed (E). The cutting plate is once again removed and the whole fringe under it (till the edge CD of plate ABCD) is again given the sharp cut (Figure 6.7).

This whole portion (N) is finally weighed. Thus, we have three different weighings as follows::

C = Combed-out portion
E = Extended portion and
N = Normal portion

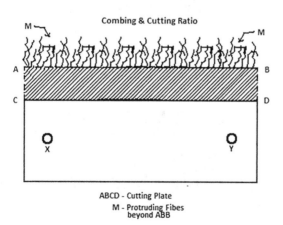

FIGURE 6.4   Cutting plate and protruding fibres: An intermediate step in proceeding for combing and cutting ratios.

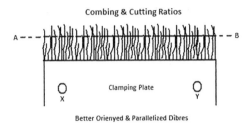

FIGURE 6.5 Parallelized fibres:[2,9] The brushing makes the fibres lie all perpendicular to the plate and parallel to themselves.

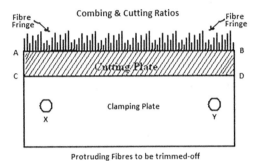

FIGURE 6.6 Light brushing: When the front portion of the fringe is lightly brushed, sweepings attached to brush bristles are removed and saved. The brushing makes all the front edges of the fibres parallel.

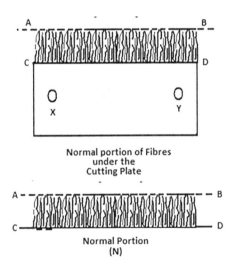

FIGURE 6.7 Trimming of fringe:[2,9] After trimming the extended fibres beyond the cutting plate, all fibres under it are finally cut (Fibres N).

The various ratios, according to the definitions, are calculated as follows:

Combing Ratio = C / (E + N)
Orientation Ratio = E / N (Cutting Ratio)
Orientation Index = [1 – E / N] × 100

It will be clear that the combed-out portion for a continuous filament will be zero; whereas, for material with all the fibres of the same length, the combed-out portion (C) will depend on fibre length (L) and the width of the cutting plate. It is assumed that all the fibres are parallel, uniform and equally represented in all the sections of the three weighings mentioned earlier. Therefore, when the width of the cutting plate = L (the fibre length itself), half the fibres in the section under the cutting plate, will be combed-out. The combing ratio in this case will be 1.0. As the width of the cutting plate is reduced, the combed-out portion also would reduce and hence the combing ratio.

For cotton fibres, however, there is a great variation in the length. Therefore, when the cutting plate is reduced in width to accommodate shorter-length fibres, not only the portion combed-out reduces but the extended portion (E) also becomes very small. The accurate measurement of this portion (E) thus becomes quite difficult. Also, with a smaller cutting plate width, the probability of cutting bent or hooked fibres is increased. In such cases, these fibres after cutting would appear as two parallel fibres and this affects the significance of the measurements. Therefore, a plate of a certain minimum width (usually 0.5") is commonly used. Even then, for a still shorter staple length, it may be necessary to use a still smaller width of cutting plate so as to have a sizable proportion of weighings.

If all the fibres in the samples are parallel, then E = 0 and hence the orientation index will be equal to 100 [Orientation Index = (1 – E/N) × 100]. The sample material with a lower order of parallelization would exhibit a lower orientation index number.

6.1.5.2   Importance of Combing and Cutting Ratios and Their Use[9]

The combing ratio is seriously affected by the crisscross arrangement of the fibres i.e., when a great proportion of fibres in the material is disposed of at random. This state of fibres in the material is easily seen in the card web, blow room lap or even bale cotton. When the proportion of such fibres is more, higher values of combing ratios (i.e., higher combed-out portions due to the high proportion of crisscrossed fibres) are obtained. In such cases, the orientation index has little meaning. Thus, the higher values of combing ratios indicate a very high disorder of fibre arrangement in the material. As a result, in such cases, the combing ratios when taken at right angles to each other are nearly equal. Therefore, when the values of combing ratios are higher (say greater than 1.5), there is not much significance to the values of orientation indices. Thus, the combing ratio is used as a rough index of randomization of fibre orientation; whereas the orientation index is best suited for the material with reasonably well-aligned fibre arrangement (i.e., drawing frame sliver, comber sliver, roving, etc.).

The method of finding the combing and cutting ratio has wide applications. The research workers[12] have found that these ratios in leading and trailing direction (w.r.t. direction of delivery) are greatly affected by the number and nature of leading and trailing hooks. Thus, the fibres with the trailing hooks (opposite to the direction of the arrow in Figure 6.8) greatly influence the combed-out portion. The cutting ratio is also a good measure of leading or trailing hooks depending upon whether they are measured in direction of feed or opposite to it. It is thus possible to use these two ratios to determine the effect of speeds, settings, draft and drafting system, type of process and parameters involved, in the fibre orientation.

In yet another research study, the effect of the drafting process on the straightening of hooks was investigated and it was found that drafting can more easily straighten out the trailing hooks. This was also found to be applicable in the final spinning of yarn on the ring frame.

Thus, the quality of the yarn is found to be better when the majority of the hooks (as at the card web) are fed to the ring frame in the trailing direction. It is now accepted that there should be an odd number of passages in between the card and the ring frame for carded mixings.

The possibilities of fibre loss in combing are basically due to two factors: (a) the chances of fibres being removed during combing by cylinder and (b) fibre breakage. The above indices can be useful in giving a fair idea about the effect of pre-combed processes in reducing fibre crisscrossing and entanglement, both of which govern the above two factors.

In recent fibrograph techniques, which are still faster than finding combing and cutting ratios, the span lengths are found out. It is observed that the span lengths (66.7%, 50% and 2.5%) are distinctly more in major hook directions.

6.1.6 Fibre Loss in Combing and Control of Comber Noil[1,4]

6.1.6.1 Fibre Loss: The Possibility[1]

It is possible to judge what the comber does in terms of the short-fibre extraction. As mentioned earlier, the extraction of short fibres always involves long-fibre loss.

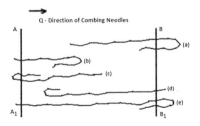

FIGURE 6.8 Direction of combing:[10,11] If the brushing direction matches the direction of the hooks, those hooks get straightened out.

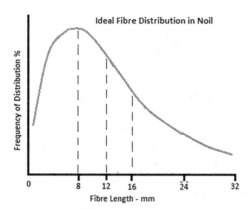

FIGURE 6.9   Ideal noil distribution:[10,11] The ideal noil should contain a maximum proportion of fibres shorter than the predetermined length and an almost insignificant representation of longer fibres.

A graph (Figure 6.9) when plotted for the fibre distribution of noil can reveal the proportion of short fibres as against that of longer fibres. The ideal graph is always skewed so that the peak is seen around a shorter fibre length.

As seen in the graph, there is a sharp fall, tapering after the peak around 8 mm. The fall is still marked up to 12–16 mm. However, thereafter, the fall rate goes on slowing down. In this case, if it is assumed that the boundary length is (say) 12 mm, then it can be concluded that the short-fibre extraction is very effective. The extended region of the graph beyond 16 mm is a clear indication of long-fibre loss. The performance of the comber, with such a type of graph, can be depicted by comparing the area under the curve – one, up to (say) 12 or 16 mm and the other beyond. The ratios of these respective areas to the total area under the curve could indicate the performance of the comber in terms of short-fibre removal and long-fibre loss.

The removal of short fibres, in general, during combing is governed by the fibre distribution curve, boundary settings, intensity of half-lap action and control over fibre acceleration during detachment. Under the ideal condition, in forward feed, all the fibres shorter than (D – f) would go into noil and those longer than this are expected to form a sliver.

Due to the typical nature of the fibre distribution and the disorder in fibre orientation (non-parallelization and hook formation), many longer fibres are treated as shorter-length fibres and hence are removed as noil. Further, the half-lap needles are not able to remove all those fibres which are not gripped by the nippers. Also, the top comb cannot prevent the acceleration of all those fibres which are short enough and which are not gripped by detaching rollers. These shorter-length fibres ultimately go to form the sliver.

A fibre is likely to break during a combing operation for several reasons. It may be due to fibre disorder or the fibres may be weak at the point of the break. The excessive friction between a fibre and the needles (cylinder half-lap or top comb needles) may also be the reason for the break. Yet another cause is that the cylinder needles are likely to force themselves while penetrating through the lap fringe. In a similar manner, the top comb needles may pose too much interference for the smooth passage of the fringe being detached. In this case, if the fibre is longer, its chances of breaking during combing also increase. There are several possibilities: (1) when a fibre breaks during combing, the broken part is obviously extracted as noil and (2) if it breaks during detachment and the broken part is a tailing portion, it will be invariably stopped by the top comb. However, if the broken part is a leading portion and it is being pulled by detaching roller, it will form a sliver. This is because; it is highly unlikely that both the broken portions of fibres would go to waste. As a result of fibre breakage, the sliver will be short of such longer fibres and at the same time, may contain a sizable proportion of short fibres. Yet another possibility is that the broken part may be taken away as noil. In this case, the noil may simply increase.

When the step gauge is widened, the cylinder needles penetrate deeper into the lap fringe. Also, there are longer boundary lengths. Therefore, the possibility of removal of short fibres is increased when the wider step gauge is used. Normally, a top comb is expected to restrict the unusual acceleration of short fibres. However, in many such instances, its action is very likely to lead to longer fibre breakage. When the majority of the hooks, are presented to the comber in a trailing direction, it also results in long-fibre breakage. This is because; during detachment, the fibre acceleration increases; whereas, the top comb, to a certain extent, fails to prevent their premature advance. In the subsequent combing cycle, these fibres are likely to be extracted during combing by the cylinder.

During detachment, a few longer fibres are also prematurely dragged ahead and out of their turn. These simply accumulate behind the top comb. They are thus, likely to lose their position under the nipper bite in the subsequent combing cycle and eventually would get removed by cylinder needles. The combing of the fringe brought back by the detaching roller for the piecing-up operation must also be correctly set and timed; otherwise, there is a likelihood of back combing.

When there is a high proportion of fibre disorder, it increases friction during combing. The cylinder needles have to forcibly pierce through these entanglements to carry out combing. The tension arising out of this forced movement may lead to fibre breakage.

The presence of leading hooks can also cause fibre breakage. Considering the nipper grip, as regards the pressure on top nippers, the effect of spring pressure (Figure 6.10) is very prominent at A. This means that a fibre held at this point will be

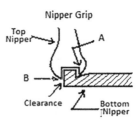

FIGURE 6.10 Fibre position and nipper grip:[1,2] The top nipper is only perfect at the top flat portion of the bottom nipper. At the front edge of the top nipper, there forms a very thin gap, where the grip is a little loose.

firmly gripped. Thus, it is highly unlikely that this fibre would slip. As against this, comparatively less pressure is experienced at B. This is owing to a small clearance (red shaded portion) between the tip of the top nipper and the front side edge of the bottom plate. Thus, when a fibre having a leading hook is held at the nippers with both of its limbs caught by the nipper grip, its ultimate fate will depend on whether the individual limb is caught at A, between A and B or at B. If the two fibre limbs are very firmly gripped by the nippers, then the breakage is almost certain. The cylinder needles while piercing through the hooked limbs would certainly break it. The value of the feed length also decides the ultimate fate of the fibres e.g., if the hooked portion is less than its distance from the cylinder needles, it may not be removed by the needles. During feeding, if the length fed is (say 'f') and the distance of this hooked portion from cylinder needles is (say 'a'), then the fate of this hooked fibre portion will depend upon:

Hooked fibre portion > (a + f) – it will be removed either in the first or second combing.

Hooked portion < (a + f) – it may be saved as useful fibre.

However, it may be proved experimentally that even if the hooked portion is greater than the critical value (a + f), it may not be necessarily extracted by the cylinder needles. This is because the normal needle spacing is wide enough to permit the catching of all such hooks. Second, it is not possible for the needles to penetrate all these hooked portions in the fringe across the width. In addition to this, the portion of these hooked fibre exceeding its critical value (a + f) is significantly low.

The probability of fibre loss in both very short and very long-fibre regions is very high. In the case of the longer ones, it is due to fibre breakage and trailing hooks; whereas, in the case of the shorter ones, it is due to fractionation. The loss is comparatively lower in the middle group.

6.1.6.2   Control of Comber Noil[4,13]

Owing to increasing prices of seeds, fertilizers, and chemical sprays for controlling pests, cotton prices are increasing. Therefore, it has become more important to save unnecessary extraction of waste at every point during the spinning process. In blow room and carding, though it is essential to remove trash and foreign matter from bale cotton, the improved technology has helped in reducing good fibre loss by modifying the waste extraction methods. As the improvement in quality at combing depends upon the level of noil extraction, a judicious selection of waste level at combing has become more important.

The normal noil level in the mill varies from 10–12% to 18–20% for medium and fine count respectively. Whereas a far more strict control is kept both in the blow room and carding, the actual values of waste% at comber for any count vary from mill to mill. The approach of such selection seems to be almost very casual. In some mills, a minimum level of just 10% noil is extracted. For similar mixing, the other mill sets its level at 12%. In such cases, some improvement is wished and some is obtained. How much, is never thought of and therefore never related.

It is important to note the nature of fibre distribution as revealed by the unique Baer sorter diagram. The level of improvement is very closely related to the nature of this diagram. It governs the quality improvement of combed material. Also, it is equally important to note the actual values of the settings used to arrive at a particular level of waste extraction.

The fitters in the mill section are not adequately educated and they change the settings very freely at their will. This, they do, when the supervisor informs them about the actual level of noil extraction and the one intended. Some of the settings like top comb depth or its timing, detachment setting, feed roller clicking timing or distance of half lap from nipper are often changed by the fitter, to adjust head-to-head variations. In a normal sense, these settings are decided right in the beginning for a particular value of the effective length of cotton and on the basis of the value of the feed length.

It is absolutely essential that when choosing a particular noil level, there should be minimal changes in the major settings. When the waste level is required to be changed by a large margin, only the step gauge should be used; whereas for minor changes (less than 1%), changes should be made only with top comb timing or its setting., provided that its interference does not spoil the web quality.

The important settings in combing that decide the yarn quality are: (1) distance between cushion plate and half lap – called as leaf gauge. This basically controls the needle penetration into the lap fringe and hence affects the fractionation. (2) The top comb setting (its depth, timing and its distance from detaching roller – all control the top comb influence in restricting the short fibres from going to the sliver. These settings are required to be optimized as they have an influence on yarn quality. (3) The nipper closure timing and the spring pressure on the top nipper control the effectiveness of the nipper bite during cylinder combing. If the timing for nipper closure, top nipper pressure and the distance of the nippers from the detaching roller, are carefully looked into, unnecessary good fibre loss can be avoided. The distance between the feed roller and the nipper depends upon the staple length processed. During detachment, this setting facilitates free movement of the fibres from the uncombed part of the lap. The normal allowance given, while carrying out this setting is 2/8" to 3/8" over the effective length. Indirectly, this setting has a relation with the step gauge. With a wider step gauge, a closer feed roller setting is necessary. If anything goes wrong in maintaining this relationship, the uncombed portions appear in the combed web.

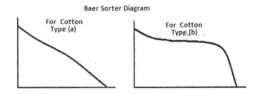

FIGURE 6.11 (a) and (b) Baer sorter diagram:[2,11] When the waste extracted at comber is increased, the level of improvement in the yarn quality depends basically on the nature of Baer sorter diagram, with (b) the improvement is very much limited after a certain waste level, with (a), however, the improvement is seen even beyond this waste level.

6.1.6.3 Timings[1]

The two most important events – the timing of combing by cylinder and the start of forward motion of detaching roller and nippers need to be carefully looked into and should be well taken care of. It is already mentioned in the earlier chapters that the various events in a combing cycle, take place at précised index numbers. All these must be synchronized to obtain efficient combing and piecing operations. Too late or too early timing of detaching rollers with respect to forward movement of nippers, often leads to prominent piecing wave in the comber web. In such cases, either a thin place (Figure 6.12 (a)) or a thick place (Figure 6.12 (b)) is formed at every piecing joint. Whereas a late timing of detaching rollers leads to cutting of the fibre front-ends, too early timing of detaching rollers results in their earlier backward motion – a possibility of back combing. This is because, it brings back the already combed fringe to possibly get exposed to the last few rows of the half lap.

Thus, some long-fibre loss is evident. With a little late timing, this possibly can be totally avoided. Especially with long-staple cotton, it is necessary to obtain an appropriate detaching roller timing, to get satisfactory piecing without curling of web and avoid any fibre loss.

When the needles on the half lap are rough, bent or damaged, they tend to pluck the long fibres. It is always advisable to repair or replace the damaged needles.

The settings of a comber critically affect the yarn quality. With six or eight heads, it becomes very essential to keep the performance of these individual heads at a uniform level. While checking for head-to-head variations, many a time, the fitters try to adjust and correct the variation in the respective heads. This is done by altering the settings related to that head. The real cause is, therefore, never found. Thus, the remedy can be worse than the disease. Also, there is a common belief that comber settings get disturbed owing to the oscillatory motion of the nipper body. If the mechanical condition of the comber is sound, the maintenance schedules are

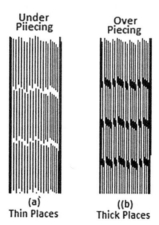

FIGURE 6.12   Piecing waves:[2,11] The adjusting superimposition of the two fringes is always very important. If this superimposition is less or more, there will be a pronounced piecing wave.

uniformly spaced and are properly carried out; and if good habits are developed in fitters and tenters, the settings which are done periodically are not likely to get disturbed. Therefore, unless it is absolutely necessary, the normal periodic settings, once done, should never be tampered with.

REFERENCES

1. *Manual of Cotton Spinning: "Draw Frames, Comber & Speed Frames": Frank Charnley*, The Textile Institute Manchester, Butterworths, 1964
2. *Elements of Cotton Spinning: Combing - Dr. A.R. Khare*, Sai Publication
3. *Spinning: Drawing, Combing & Roving: Book of Papers: Dr. R. Chattopadhyay, Dr. R.S. Rangasamy: NCUTE Programme Series*, 1999
4. Comber Noil Control - TRJ. 1959, 742
5. Combing Cotton/Viscose Blend: K.I.Badalov, Tech. of Textile Industry U.S.S.R., 1962
6. Fractionating Efficiency: R.G.Owalekar, Proc. 13th Tech. Conference, ATIRA, BTRA, SITRA, 1965
7. Fractionating Efficiency: M.S.Parthsarathy, Proc. 13th Tech. Conference, ATIRA, BTRA, SITRA, 1965
8. Fractionating Efficiency – Simpson & Ruppenicker, *Text Bulletin*, 1960
9. Measurement of Fibre Orientation – Charles Lindsley, *T.R.J.* 1951
10. Fibre Orientation: Wakankar etal, *Textile Research Journal*, November, 1961
11. *Technology of Short Staple Spinning: W. Klein*, Textile Institute Manual of Textile Technology
12. *Spun Yarn Technology – Eric Oxtoby*, Butterworth Publication, 1987
13. *Process Control in Spinning, A.R.Garde & T.A. Subramanian, ATIRA Silver Jubilee Monographs*, ATIRA Publications, 1974

7 Faults in Comber

7.1 FAULTS IN COMBER[1,2]

Mechanical faults in the comber result in excessive variation in the quality of the web produced. It is, therefore, essential to check the machine settings and timings along with the condition of working parts and take corrective action.

7.1.1 Damage to Half-Lap and Top Comb

Both the half-lap and top comb are very important parts in the comber and are considered the heart of the combing process. The saving in re-needling or repairing does more harm in the long run and hence the expenditure on these should be incurred as and when the needles are found damaged. Thus, cleaning of the half-lap or straightening of the bent needles should be done promptly. The double-end piecing of the lap normally leads to the development of sudden stresses on the needles. Hence the tenter should be trained for carrying out correct lap piecing of an old and a fresh lap. In the latest modern combers, automatic lap piecing is provided to eliminate the fault of double-end lap feeding. Along with this, it is essential to replace excessively worn-out brushes. This is because, as the bristles become shorter in length, they lose their flexibility. As a result, not only the cleaning of needles is affected but their rigidness also puts additional pressure on the needles. With new brushes, loading on the half-lap is easily avoided and at the same time, the possibility of needle damage is much reduced.

One of the reasons for the damage to the top comb is its faulty setting with detaching rollers. Too close a setting, makes the top comb touch the top roller surface occasionally. Here too, along with the needles, it also spoils the top roller covering. Further, when the top comb is allowed to enter the fringe too deeply, the resistance that its needles experience during detachment is likely to cause some damage to the needles.

7.1.2 Uneven and Inadequate Nipper Grip/Feed Roller Grip/Detaching Roller Grip

It is very important to have the uniformity of nipper bite across its width and feed roller and detaching roller grip across their length. As for the nippers, this can be checked by inserting a thin paper strip between the two nippers.

When the nippers are closed, the paper should be evenly gripped when moved from one end to the other. The same also can be tried with a feed roller. This is because, both these are spring-loaded. In these cases, it is advisable to check the

DOI: 10.1201/9780429486555-7

position and the condition of these springs. With insufficient nipper grip, there is a possibility of long fibre loss; whereas with inadequate feed roller grip, some uncombed portion of the lap fleece would be allowed to escape into the comber web.

The loading on the detaching roller is directly related to the efficiency with which detachment takes place. Inadequate pressure, especially on the back top detaching roller results in long fibre loss and also leads to back-end combing.

7.1.3 Curling of Fibres in the Web

This is a typical defect that occurs when the already combed fibre fringe is withdrawn back for piecing. This fringe follows the curved surface of detaching rollers and hence leads to wild piecing.

The common reasons for such defects are: faulty detaching roller covers – either spoiled or not cleaned, too high or too low humidity and faulty passage of air currents generated by the aspirator i.e., the air entering the detaching zone and passing on further to the aspirator.

7.1.4 Holes in the Web

These are mainly due to disturbances in the web caused by air leakages. A faulty setting between the air strip and the brush, leakages in the air seal or worn-out brush bristles – are the common reasons for this defect.

7.1.5 Uncombed Portion in the Web

Some of the portions of the lap, without being properly exposed to the cylinder combing actions, escape to form the web. This usually occurs at the time of detaching fleece. Uneven laps, ineffective feed roller pressure and too wide setting of the feed roller from the nipper bite are common reasons.

7.1.6 Plucking

This is commonly experienced with long staple fibres where the half-lap needles enter the fringe with sudden impact carrying away the long fibres. The probable reasons are – variation in the lap weight – along and across the length, poor fibre parallelization, ineffective nipper grip, rusty-rough or bent needles on the half-lap.

7.1.7 Cutting Across

In the web, thin and thick places appear regularly; so the web appears to be cut at these places.

The fault may be due to bad laps prepared at the lap machine, incorrect roller settings, excessive drafts or roller slip. With incorrect timing of detaching cam, the faulty piecing of the web can also result in thin–thick places appearing at regular intervals.

Wrong position of the top comb – too deep placement of comb into the web and/ or very close distance of comb from the detaching roller also lead to its excessive interference during detachment and cut in the web.

7.1.8 WEB NOT COMING THROUGH HEADS

During the normal working of the comber, it is sometimes observed that there is no web or sliver coming from one or more heads. At times, the lap at the back gets exhausted or there is lapping around detaching rollers. Both can be detected by a vigilant worker and the time lost in repairing the damage to the product can be kept to a minimum.

However, in some cases, it is necessary to observe whether there is excessive fibre loss on any of the heads where most of the fibres are taken away by the cylinder. This may be due to the ineffective bite of the nippers, foreign matter wedged in between the nipper bites or even a faulty detaching roller. In the last case, the detaching rollers are required to be lubricated to ensure their smooth running. In some other cases, the pawl engaging the feed ratchet gets disengaged or the ratchet itself becomes loose. This stops the feed to the corresponding comber head.

7.1.9 LAP RUNNING SLACK BETWEEN LAP ROLLER AND FEED ROLLER

This is basically due to improper co-ordination in speeds of lap and feed rollers. Either the former is feeding excess of lap or the latter is taking a little less. Even, as stated earlier, the pawl on the feed roller may be disengaged or the feed roller ratchet may have become loose on its shaft.

7.1.10 LAPPING ON DETACHING ROLLERS

The defect is mainly due to wrong atmospheric conditions (higher R.H.) or bad condition of detaching roller covers. In the second case – oil on the roller cots, badly worn-out cots requiring varnishing, dirt sticking on the cots and hence requiring cleaning or improper roller clearers – are some common reasons.

7.1.11 FLOCKING OF WASTE ON CYLINDER NEEDLES

One of the important operations after combing by cylinder is the stripping of cylinder needles by the revolving brush. But owing to several reasons, this is not done effectively and efficiently. One or more brushes get loose on the shaft or sometimes uneven wearing-out of the brush bristles results in a gap between the bristles and the cylinder needles.

The air leakage through the aspirator is yet another problem which reduces the strength of the suction. The deposition of fibres on the brush continues, but the waste on the bristles is not drawn by the suction from the aspirator. Hence, the respective brush loaded with fibre noil is no longer able to strip the cylinder needles. As for the suction, the function of the air strip placed over the brush is very important. The air suction controls the effectiveness of the stripping action. If the brushes are well cleaned by the suction they, in turn, strip the needles efficiently. On some modern Rieter combers, 'a slow motion' for the cylinder is provided at intervals, wherein except the brush shaft, all other parts are made to periodically run at slow speed. This allows a regular and efficient cleaning of needles.

When the cylinder needling (half-lap or UniComb) is found damaged, the waste frequently gets accumulated on and around this damaged portion. This leads to the flocking of waste.

7.1.12   EXCESSIVE LOSS OF LONG FIBRES[1]

During the working of the comber, it is important to carefully observe, from time to time, the nature and the feel of the noil collected on the aspirator drum. When the noil contains excessive whitish long fibre tufts, they are easily noticeable. Even the Baer sorter diagram of the noil taken in such cases can easily reveal this.

Improper nipper bite has a direct influence over the loss of long fibres during combing. Sometimes, a foreign matter gets wedged between the nipper jaws. This invariably leads to uneven gripping of the lap fringe. Even the irregular lap across the width leads to the same result.

Bad covering of detaching roller leads to ineffective gripping of fibres during detachment of the fibre fringe. The damaged needles of the top comb or its wrong setting increase its interference. This causes serious disturbances during the detachment of the fibre fringe. The longer fibres are unnecessarily withheld by such disturbances and fail to pass on to the detaching rollers. In the subsequent combing cycle, these fibres are simply removed during the cylinder combing operation.

7.1.13   IRREGULAR DRAW-BOX SLIVER[1,2]

Normally, owing to the piecing wave, the comber head sliver becomes irregular and weak. However, apart from this, head-to-head variation in the noil, also results in non-uniformity in the hank of sliver delivered on the sliver table.

When the head sliver passes over the sliver table, there is always some rubbing over the table surface. It is here that the slivers are very likely to experience 'false draft' (non-intended draft) if the table surface is slightly rough. The tension draft between – (a) the table calender roller and detaching roller and (b) the back draw-box roller and table calender roller – have to be within limits (not more than 1.05). This is because the distances at these two controlling zones are much wider than the effective length. Any stretching in these two zones directly leads to irregular table slivers. The breakages of sliver/s occurring in the above two regions lead to a defect – 'singles and doubles'. If the worker minding the machine is not alert, these are normally not attended to immediately. This again leads to irregularity in the draw-box sliver.

The bad condition of the detaching roller covering, improper settings and weightings in the draw-box can all aggravate the situation. The use of an ordinary drafting system (conventional 4/4 drafting) leads to the 'phasing' of the piecing wave. Many comber manufacturers, with their new high-speed models, have introduced modern drafting systems such as 3/5 or 4/5. In some mills, the conventional drafting system is modified as 2/2 or 3/3 drafting with an improved weighting system on top rollers. However, the amount of drafts that can be introduced in such cases is limited. It may be noted here that this draft in the comber draw-box is not employed to improve orientation and parallelization. This is because the fibres in the comber sliver already attain a good deal of both of these qualities, again due to cylinder needle and top comb needle action. In some combers, the head slivers are divided into a group of four and bi-coiling is used. In this case, the draft involved for each sliver group is only around four. In a Whitin comber a 4/5 system is used; whereas, in Rieter's E-86 comber, a 3/3 system is employed.

7.2   POST-COMB DRAWING FRAMES[3]

The necessity of joining the two fringes during piecing makes the comber sliver very irregular. According to Nutter and Slater[3] these piecing waves come into phasing in post-comb drawing passage. Hence, the full benefit of doublings in the postcomb drawing is not realized. As stated earlier, the concentration of draft in only one zone has been proven to control this phasing. In most of the post-comb draw frames, therefore, low break drafts are recommended. Although the 2/2 system was originally developed for the comber draw-box, it is equally effective for subsequent draw frame passages. It is found that not only this system on draw frame improves uniformity, but also helps in minimizing the irregularity at the second passage of post-comb drawing.

In the case of a coarser comber sliver, it is essential to increase the roller grip in the draw-box of the post-comb draw frame. This can be done by either increasing the roller weighting or by suitably modifying the bottom roller flutings. The width of the flutings, in this case, can be broadened to avoid overcrowding in the drafting zone. For normal hank, the roller settings are not critical and the usual plan of +5 mm and +8 mm over the staple length, in the front and the back zone respectively holds well. However, with the higher bulk, the settings are slightly wider than recommended. Earlier research work indicates that the timing of the detaching roller cam (starting of the forward motion of detaching roller at the time of piecing) has a direct influence on the regularity of post-comb drawing. The effect of a doubling of six or eight slivers does not have much influence on improving the level of irregularity.

It is also found that the range of break drafts used on post-comb drawings has little influence on the regularity of the outgoing sliver. However, most of the mills still prefer lower break drafts of not more than 1.3 on these draw frames. Excessive sliver creel breaks, especially with short and medium staple cottons, are yet another source of poor draw frame performance. These creel breaks are mainly due to two reasons: (a) poor strength of the comber sliver because of excessive parallelization and (b) higher speed of working.

To improve the strength of the comber sliver, a coarser hank is produced. This can be done by reducing the draft in the comber draw-box to not more than three to four so that a sliver of double thickness is produced. As compared to a higher draft of eight, the number of doublings in the draw-box is thus halved due to a sliver of double thickness. Similarly, the number of cans behind the post-comb drawing is also reduced by the same ratio. With this, the level of break draft may hardly have any significant influence on the regularity, particularly if the settings are maintained at their optimum level. However, for a closer front zone setting, it is advisable to have a higher break draft and vice-versa.

It has been a common practice to slow down the speeds of post-comb drawing frames by 15–20% as compared to the one used for pre-comb draw frames. The reduction in the creel height, avoiding longer extended passage from the feed can to the lifter roller of the creel, can springs in really good condition and provision of a smooth surface for combed sliver over the creel table – all help in controlling either undue stretching of sliver or their frequent breaks.

It was found that, though the sliver U% was improved by the use of the second passage of post-comb drawing, no significant improvement in the yarn properties was noticed, except for the reduction of U% at the 'inert' position of USTER. This means that medium-term variations in the yarns are reduced. As a result, this reduction positively contributes towards improved fabric appearance. In spite of this, many mills use only one post-comb drawing passage. This is because the economics of using an additional passage may not bring sizable returns.

In short, it can be summarized that:

(1) Irregularity of the comber sliver is critically influenced by the detaching roller timing
(2) The effect of doubling the comber slivers on the first passage of the drawing frame does not much help in improving its quality. This is because the irregularity of the sliver fed to the post-comb draw frame has more influence on its performance in the first post-comb passage.
(3) In general, the lower break drafts are used in these post-comb passages, as there is already maximum parallelization. However, at optimum roller settings, the value of the break draft has hardly any effect.
(4) A coarser sliver, after combining all head slivers, does not seem to adversely affect its performance on post-comb drawing. On the contrary, it reduces the creel breaks at post-comb drawing.
(5) One post-comb passage is not adequate enough to improve the regularity of the comber sliver. However, the second passage does not seem to contribute much to yarn properties except reducing medium-term variations.

REFERENCES

1. *Manual of Cotton Spinning: "Draw Frames, Comber & Speed Frames": Frank Charnley*, The Textile Institute Manchester, Butterworths, 1964
2. *Elements of Cotton Spinning: Combing: Dr. A.R. Khare*, Sai Publication
3. Nutter & Slater: Critical assessment of recent progress in technology of cotton spinning, J.T.I., 1959, p.397

8 Features of Modern Comber

8.1 FEATURES OF MODERN COMBER[1]

Over the number of years, lots of changes have taken place in structuring and redesigning the comber. Even then the basic combing operation such as lap feeding, combing by cylinder or even detachment of combed fleece and joining of the two fringes have remained the same. Thus, the general comber appearance has been more or less unchanged.

However, restructuring and redesigning of parts has enabled the removal of complicated driving mechanisms and has further helped in reducing the number of parts as well as their weight. The maintenance of the comber has been simplified; the oiling and greasing procedures are made easy by using non-oil and sintered bushes, centralized oil pumps and plastic pipe connections leading to various important rotating parts. For greasing, the nipples are provided and with the help of grease guns, the operation can be done during working.

Even for the machine operator, the workload has been considerably reduced and his mind is made free due to various electrical stop motions. Thus, his operating efficiency either for the same number of machines, or even for a little more, is improved. The basic aim of adopting these various improvements is to attain higher production rates at the comber. As compared to conventional machines, present modern machines are capable of working with very heavy laps (from 20 g/m to 70 g/m and above) at remarkably higher speeds (from 90 nips/min to more than 300 nips/min). The number of heads is increased from six to eight. Some modern machines have double-sided workings. A larger diameter coiler has enabled the use of larger cans. All this has helped to increase the production from 30 kg/shift to 60 kg/hr. However, it is equally important to note that on and above these improvements in production rates, the quality of the product – comber sliver - has been significantly improved.

8.1.1 FEED

Apart from the usual tension compensator provided to take up the slackness during the backward motion of the nipper assembly, modern machines provide an easy changeover from forward to backward feed and vice-versa. Some of the models have continuous feeding arrangements. In a Laxmi Rieter comber, when changing over from forward to backward feed, the ratchet-pawl arrangement is removed and the rotation of the feed roller is effected by a toothed segment fastened to nippers. In a Whitin comber, a separate lever guiding the pin on the ratchet assembly has been changed.

DOI: 10.1201/9780429486555-8

8.1.2 Nippers

In place of the normal driving of the nippers in traditional combers using hinged support, the two rocker arms, loosely fulcrumed around the cylinder shaft gives forward and backward movement to the nipper assembly. With this, the nippers move in a concentric arc around the cylinder. The nipper plate is also redesigned to reduce its size and weight. The pressure on the top nipper is exerted by the springs which are compressed when the nippers move back, thus exercising adequate and uniform pressure on either side of the top nipper.

8.1.2.1 PP Nippers

The PP nipper is an additional gripping nipper (Figure 8.1) provided to grip the fibres during the piecing-up operation. Normally with conventional combers, it is possible that, at the time of the entry of the top comb, the front end of some long fibres fails to reach detaching roller nip. These fibres are stopped by the top comb. However, their rear end being free, those ends move ahead prematurely with the detachment of other fibres in the fleece. In the subsequent combing cycle, these dragged rear ends are very likely to come out of normal nipper grip. Hence such fibres get removed as comber noil.

To overcome this shortcoming, an additional nipper called a PP nipper is positioned just at the back of the main nippers. When the main nippers open out during piecing-up, PP nippers are made to close and hold the rear end of the fringe positively. This prevents the dragging of many long fibres which fail to

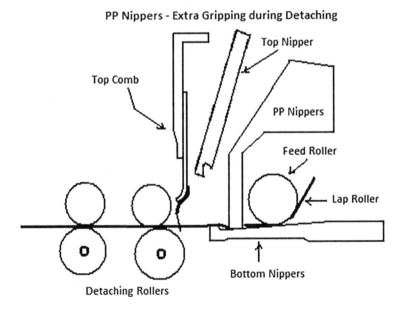

FIGURE 8.1   PP nippers:[3,4]: It gives additional support and grips the fringe, when the piecing-up operation takes place. Thus it does not allow the fibres to prematurely go ahead.

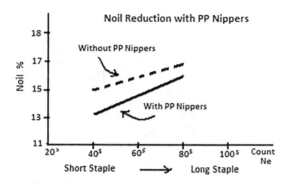

FIGURE 8.2 Noil reduction:[1,4] When PP nippers are used, there is definite saving in the noil. However, they are more useful with lower mixing (shorter effective length).

reach detaching roller nip. This is because, with PP nippers their major portion remains behind the normal nippers and during the cylinder combing they are under the positive hold of main nippers It is claimed that there is a considerable saving of good fibres to the extent of 0.5–1.5%, which otherwise would go as a waste. The graph (Figure 8.2) shows the advantage of using PP nippers demonstrating noil reduction with and without PP nippers. However, as the staple length of cotton processed increases, the reduction in the noil with PP nippers is comparatively less.

The Baer sorter diagram (Figure 8.3) clearly shows the difference between with and without PP nippers. It reveals that there is a definite saving of longer fibres. As can be seen from the nature of the Baer sorter curves, there is comparatively less proportion of long and medium-length fibres when PP nippers are used.

This means that PP nippers, in providing an extra grip during detachment, do help in saving the longer fibres. As mentioned earlier, this is because these additional nippers control the unwanted dragging of longer fibres during detaching.

8.1.2.2 Modern Nipper Assembly[2-4]

In a high-speed comber, the designing of the nipper framing becomes very important. During a combing cycle, the nippers accelerate and slow down twice in each cycle. With a speed of more than 300 nips per minute, this turns out to be almost five times per second. The nippers are, therefore, required to be very light in weight and hence are made of aluminium. Further, as the heavy laps are processed, the lap sheet is almost double in thickness.

The nippers are required to grip this thick sheet very firmly (Figure 8.4). In modern combers, the upper plate is stiff whereas, the bottom plate is slightly springy. The pivot axis holds both the nippers. Thus, the nippers can be raised or lowered around this axis. There are two springs, one each on either side to exert pressure on the top nipper during the cylinder combing operation. These springs hold the lap in between their bites very firmly. The nipper bites are specially designed (Figure 8.4 [A]) to give a more firm grip on the lap sheet held between them. The nose (N) is designed

Saving in Comber Noil with PP Nippers

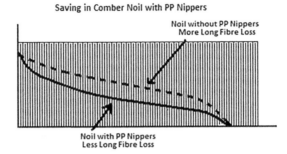

Noil without PP Nippers
More Long Fibre Loss

Noil with PP Nippers
Less Long Fibre Loss

FIGURE 8.3 Baer sorter diagram:[1,4] When the studies were made on the nature of the comber noil, it was revealed that using PP nippers certainly avoids losing longer fibre in the noil.

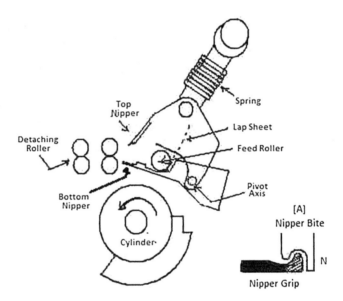

FIGURE 8.4 Light nipper assembly:[1,2] The whole nipper assembly oscillates at very high speeds. This causes a lot of vibrations and jerks and leads to excessive wear and tear of the parts. With lighter nipper assembly, this is substantially reduced.

to press the fringe down. Thus, the fringe is forced downwards and therefore, cannot escape the combing action of the cylinder needles.

8.1.2.3 Movement of the Nippers[2,4]

Two pivot levers, one each on either side support the bottom nipper plate (Figure 8.5). These pivot levers are loosely held around the cylinder shaft and are also supported by the two swing arms which are screwed to the nipper shaft. The connection from the nipper shaft to point P is via a link (swing arm) which, along with the nipper

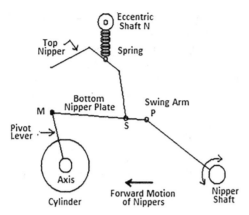

FIGURE 8.5 Nipper forward movement:[1,2] As against a separate mechanism provided for opening and closing of the nippers in conventional combers, a very simplified mechanism is provided to open the nippers.

shaft movement (in the direction of arrows) therefore oscillates to and fro, thus pushing another linkage PM forward and backward. In a full combing cycle, this motion of linkage PM actuates the forward and backward motion of the nippers. The most forward position of the nipper plate is decided by the setting between the nipper and detaching roller. The top nipper is supported at point S and is suspended from the eccentric shaft by means of powerful springs.

This eccentric plays an important role when the nippers are being closed. Thus, when the nippers are withdrawn back, they need to be closed (Figure 8.6). It is very important here, that when the nippers are being closed, there should not be any suddenness in their action; otherwise, in every cycle, there would be a sudden impact of top nippers on bottom nippers. This would invariably lead to wear and tear on both the nippers. It is here that the eccentric plays an important role. It eliminates this

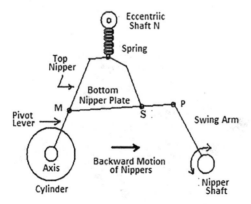

FIGURE 8.6 Nipper backward movement:[1,2] A similar simplified mechanism operates the backward movement of the nippers. Here an eccentric plays an important role in avoiding any jerk when the nippers close.

sudden jerky closure and makes the top nipper gently close on the bottom nipper, with the lap in between. The springs further ensure that the pressure of the top nipper on the lap fringe is adequately strong to hold the lap firmly.

It may also be seen that when the link PM is pushed forward, point S is also pushed ahead. This automatically opens the top nipper. Obviously, it happens during the forward motion of the nippers, when the piecing-up operation is due.

8.1.3 Cylinder Needling[2,5]

The conventional combers used half-lap with 20 rows. The earlier versions of modern combers reduced the number of rows from 20 to 17. In present days, almost all modern combers use 'UniComb' or 'Hi-Comb' or 'circular comb' segments (Figure 8.7).

They are equipped with saw-tooth metallic wires and are fitted exactly in the same place where the conventional half-lap was mounted.

A lot of developments have taken place in UniComb in terms of different characteristics. The circumference of the comb over the cylinder also varies from 75° to 110°. In general, the larger the angle the more the spread of wire points. However, the best choice depends on the type of cotton used. In any combing operation, the direction of the motions of the combing segment and that of the nippers is very crucial. The action is much more effective when the two are opposite.

[A, (B1-B2), (C1, C2, C3) and (D1, D2, D3, D4)] – Segment Strips

Another striking development is that, as against the single segment used in the early UniComb models, the whole saw-tooth segment in the latest combers is divided into more than one segment strip (Figure 8.8). Each subsequent segment, in the opposite direction of cylinder rotation, has a progressively finer and denser population of saw-tooth points. This avoids suddenness in penetration of the UniComb half-lap and hence gives improved combing action. So also, the total combing points on the whole UniComb segment are varied. This again is related to basically two fibre characteristics – effective length and micronaire. For higher micronaire and lower effective

FIGURE 8.7 Modern half-lap (CircularComb): The saw-tooth wire segment mounted on the Rieter's E-86 comber is very strong and sturdy and they have a long life.

Uni-Comb segments divided with Varying Saw-Tooth Point Density

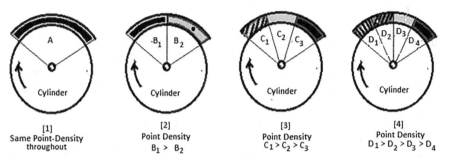

FIGURE 8.8 UniComb segments:[3,4] In place of the whole segment being made of saw tooth with the same point density all over, the modern UniCombs have the total UniComb segment divided into smaller sections with progressively increasing the saw-tooth point density.

length, the total combing points are comparatively lower (around 13,000); whereas for fine cottons with lower micronaire and higher effective length, the total combing points could be as high as (around 24,000). For lower effective length, the UniComb with two segments is preferred; whereas for longer-length cottons, a UniComb with as many as four segments is available.

8.1.4 TOP COMB[2-4]

It is carried by an additional bracket mounted on the nipper framing. Thus, the comb moves along with the nipper assembly. The circular path of the nipper assembly ensures that the top comb automatically comes into operation (Figure 8.9 (a)) when the nippers move forward and are close to detaching rollers. This avoids any additional motion to be given to the top comb.

In modern comber the top comb is detachable. The depth of the top comb into the fringe can be altered by adjusting it with the help of securing a screw (Figure 8.9 (b)).

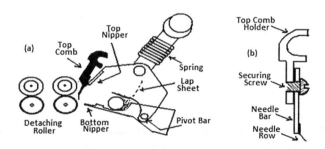

FIGURE 8.9 Top comb:[2-4] On all modern combers, the nippers move concentrically over the cylinder. Thus, the top comb automatically moves into action when the nippers move forward and closer to the detaching roller.

So also, its distance from the detaching rollers can be adjusted. Like conventional combers, the top comb on modern combers also finds its position in between the nippers and the detaching rollers. The top comb holder carries a needle bar to which the fine needles of the top comb are attached. The needles are flat and slightly bent at the bottom. They are soldered to the needle bar. The top comb is carried by the bottom nipper plate so that it swings along with the nipper movement. However, unlike conventional comber, the top comb does not have any up-and-down sliding motion. Owing to the swinging of the nippers concentric to the cylinder, the top comb gets automatically withdrawn from the web when the nippers move back. Similarly, when the nippers move forward, there again, owing to the concentric movement of the nippers, the top comb is automatically brought in the path of the fringe being detached. Avoiding the traditional up-and-down movement of the top comb reduces additional mechanisms and eliminates friction and vibrations.

8.1.5 DETACHING ROLLERS

Unlike conventional combers, there is no swinging of the back top detaching roller. The forward and backward rotation of bottom detaching rollers is achieved through 'sun and planet' or any other modern type of differential mechanism. With this, therefore, detaching rollers are hardly stationary. They slowly move forward to deliver the fringe and move backward at a faster speed to bring back the already combed fringe for piecing.

Special helical flutings are made on bottom steel detaching rollers. The top roller cots are specially designed to take care of both their hardness and anti-static nature. This, on one hand, improves the detaching roller grip and at the same time reduces lapping tendencies. The weighting system on the top detaching rollers is of spring type with the saddle arrangement or pneumatic type with special weighting hooks. The weight-releasing handles or switches are provided in the case of cleaning of top detaching roller. Some combers have detaching rollers of smaller diameters. This facilitates the combing of short-staple cottons and thus makes the machine more versatile.

8.1.5.1 Detaching Roller Drive

One of the limitations of increasing the production of conventional combers was that the various cams were employed in driving different parts. The cams limit the speed of the machine, as at higher speeds, they set the vibrations, which in turn, lead to increased wear and tear of the parts. The vibrations also deteriorate the quality performance of the various operations involved in the machine's functioning.

In one of the modern mechanisms, the link is used with larger needle bearings. They give forward and backward motions to detaching rollers. The links work on the basis of bell crank levers, where the motions are added or subtracted depending upon the resultant motion to be imparted to detaching rollers. The mechanism is very simple to operate and as there are no heavy cams and bowls involved, the drive to the detaching roller is smooth with much less friction.

8.1.6   WEB CONDENSATION AND SLIVER TABLE

As stated earlier, the trumpet in front of detaching roller to collect the web is situated asymmetrically. This reduces all the ill effects of the piecing wave. The table calendar rollers are heavy and have coarser flutes for better gripping action. As there are eight heads, the slivers have to pass over a longer distance over the surface of the sliver table. The table surface is, therefore made quite smooth by giving a special plating. This helps the head sliver to pass over the table without any stretching. An additional calender roller is provided halfway along the table length to avoid the dragging of slivers.

8.1.7   DRAW-BOX

The common drafting system used in the draw-box is of the type – 2/2, 3/4, 4/5 or 5/4. The top rollers are weighted with an over-arm weighting system or with pneumatic weighting. The gauge is provided in the latter case to enable reading of the correct pressure. The bottom rollers have special flutes and are equipped with needle-bearing bushes. The slivers are fed to the coiler calender rollers in the usual way, i.e., four slivers are grouped together and two such emerging sliver groups are coiled separately with either a twin coiling or bi-coiling system. The draw-box draft is reduced to deliver a coarser sliver. This reduces the possibility of its stretching in a subsequent process.

8.1.7.1   Typical 5/4 and 2/2 Drafting Systems

Different drafting systems are used in the comber draw-box. Apart from precision in controlling

the shorter fibres, the drafting capacity of the system also becomes important. One such typical system is shown in Figures 8.10 and 8.11. The 5/4 drafting system can give a draft of up to eight with an eight-head comber. With a reduction in the draft, a thicker sliver can be produced. Usually, a single coiler system allows a can-changing mechanism to reduce the workload. With a 2/2 drafting (Figure 8.12), as mentioned earlier, the phasing of the piecing waves is considerably reduced as there is only a single drafting zone. However, the draft that can be employed is limited. This is taken advantage of by using a bi-coiling system, where the two slivers of the normal thickness emerge.

8.1.8   COILER

With higher production rates, the coilers are required to be redesigned to accommodate larger cans. As mentioned above, they are provided with a twin coiling, bi-coiling system or one similar to a conventional single-sliver system. The hank meters are provided to register a certain length of sliver into the can. They also register the total production in hanks at the end of every shift. On some machines, an automatic can-changing arrangement is provided. This helps to reduce the workload of the tenter who has to only put empty cans in the stock of the creel and take out automatically delivered full cans.

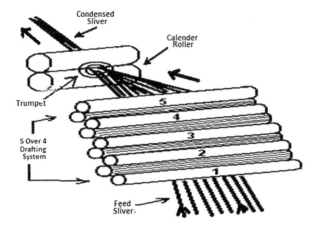

FIGURE 8.10 Five-over-four drafting:[3,4] With an odd number of top and bottom rollers, especially more top rollers than bottom rollers, the extended surface contact minimises the irregularity effect in the comber sliver.

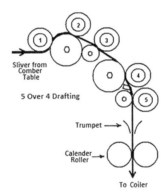

FIGURE 8.11 Five-over-four drafting:[3,4] It is capable of processing a heavier sliver and provides an excellent fibre guidance system

8.1.9 Larger Diameter of Cylinder

In some modern combers, the total covering area (Figure 8.13 (a)) for the pins or the saw-tooth points is increased.

Therefore, the larger combing surface leads to improved cylinder combing action. The needle or saw-tooth sector covers 113.5° in place of the earlier segment which only covered 86°.

Figure 8.13 (b) shows the needling of a typical modern comber and the clearance for the needle points from the bottom nipper. This is another needling arrangement which has been found to be more effective in short-fibre removal during main

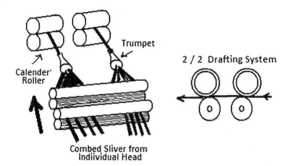

FIGURE 8.12   Two-over-two drafting:[3,4] The research work carried out on this drafting system has proved that the irregularity due to the piecing wave is minimised.

cylinder combing action. A small variation is seen in the last few needle rows. These rows remain at a constant distance from the bottom nipper. In the earlier model, this distance from bottom nippers was slightly increased.

8.1.10   Avoiding Inverse Airflow by Brush

The suction through the aspirator always demands an air supply from and around the brush section. The air needs to be drawn from this section, which is partly guided around the cylinder. In a conventional comber, a small proportion of air managed to get through the space between detaching rollers, over which, the pieced-up portion of the web was being passed. This was basically due to inverse upward airflow from the brush to the cylinder. This air used to cause disruptions in the smooth web delivery by detaching rollers. The air control pieces are used under and in between the cylinder and brush. These pieces guide the airflow without any turbulence into the passage leading to suction at the aspirator.

As seen in Figure 8.14, the airflow is also suitably diverted away from the top comb so that it does not disturb the very important piecing-up operation. This airflow is led towards the safe region around the cylinder.

Similarly, the distance between the bottom nipper and the cylinder is widened during the piecing-up operation. This again results in avoiding any turbulence.

It may be noted here that any turbulence in the region between the nippers and the detaching rollers during the time of detachment is likely to break the web which is being delivered. This is because, the turbulent air, in trying to find its way through, usually leads itself through the web being detached (torn web).

8.1.11   Nylon Filters

In conventional combers, the aspirator draws the air from the brush shaft region. The air thus drawn cleans the brush and brings an air-noil mixture onto the perforated surface of the aspirator. The separation of noil and air takes place at the perforated

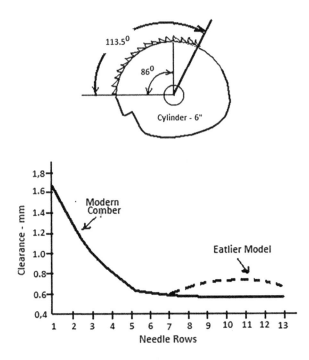

FIGURE 8.13 (a) The greater angle over which the unicomb half-lap surrounds the cylinder:[2,6] It allows a larger pinning area. And gives improved short-fibre removal. (b) Needle clearance:[2,6] The closer clearance provided by the needling arrangement for the later part of the segment endures the better combing.

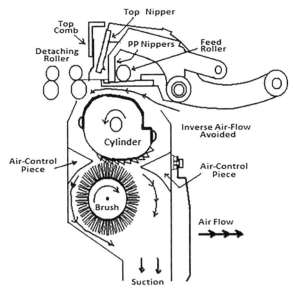

FIGURE 8.14 Avoiding inverse airflow:[2,6] The direction of rotation of the brush shaft is opposite that of a cylinder. The air thrown by the former disturbs the piecing-up operation. The air control piece provided protection from such inverse airflow.

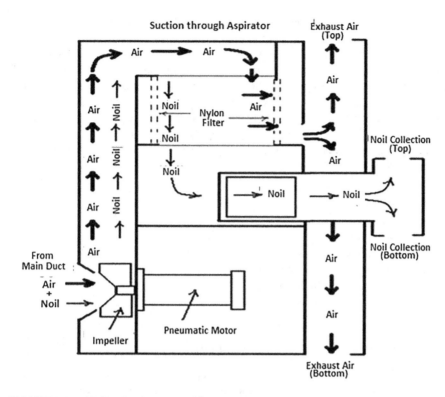

FIGURE 8.15 Noil collecting device:[2,6] The air sucked through perforations of the aspirator before being allowed to go into the surrounding atmosphere, is filtered by passing through nylon filters. This helps in maintaining the surroundings clean.

surface of the aspirator. However, the perforations do allow a small percentage of very short fibres along with the fine dust to pass through them.

Inside the aspirator, the air leading further brings this mixture into the exhaust fan region. To avoid this, the modern combers (Figure 8.15) are provided with nylon fine mesh filters which are very effective in separating the noil from the air and allow only air to be drawn into the fan-suction area.

8.1.12   HEAD-STOCK

The head-stock is placed in a closed compartment and only the main shaft 'V' pulley receiving the drive from the motor, projects out. In some cases, even the motor itself is enclosed. The detaching roller differential motion drive is placed in an air-tight compartment which is invariably immersed in an oil bath. A master gauge is provided for altering the step gauge settings on all heads simultaneously. In the Rieter comber, a slow motion arrangement is provided for driving all the parts except the brush shaft at regular intervals. This helps in the periodic cleaning of the half-lap. It thus helps in improving the combing action by the cylinder needles and reduces the neps in the web. There is a time switch provided for this operation.

Safety guards are provided for locking the doors, which enclose the machine gearing. During the running, if the doors are unknowingly opened, the machine automatically stops. The stop motions, – for lap exhaust, lapping at feed and at detaching rollers, choke-ups at trumpet, lapping and breaks at draw-box coiler – all ensure trouble-free running. Indicator lamps help by signalling the appropriate cause of machine stoppage. The starting push buttons and inching buttons are suitably located at convenient places on the machine. All these, ease the pressure from the operator's minds and help in improving his operating efficiency.

In high-speed combers, the increase in production is made possible through the following things:

1. Higher nips/min
2. Heavier lap feed
3. Higher feed length
4. More number of heads per machine
5. Better machine efficiency – bigger feed and delivery packages
6. Better operating efficiency and
7. Reduced maintenance time

8.1.13   Combing Parameters and Yarn Quality

Apart from the improved mechanical design and the ease of operations, the developments in combers offer several other advantages such as efficient combing of short and medium cottons to improve their spinning value, reduction in comber noil without adversely affecting yarn quality, increase in the productive capacity of the machines to reduce manufacturing cost, etc.

The attainment of adequate lap regularity for efficient combing and better head-to-head uniformity of comber sliver is mostly dependent on the type of drafting system employed on the pre-comb drawing and lap-making machine. Platt's 3/3 or 2/3, Whitin's 4/5 and Laxmi Rieter's 3/5 systems on draw frame are capable of suppressing irregularity patterns due to roller slip and considerably reduced drafting waves. They also restrict and control the short-term irregularity of card sliver and its subsequent amplification due to phasing from entering the comber lap.

8.1.14   Comber Lap

With given settings of comber, the waste extracted generally increases with higher lap weights. Lap prepared with a low number of doublings and draft shows a greater increase in waste percentage as the lap weight increases. The condition aggravates when a longer feed length is used. It is observed that, with poor fibre control during the lap preparation, the heavier laps tend to produce inferior yarn in terms of Count strength product (C.S.P.).

It can be seen from Figure 8.16 that the pre-comb draft bears a relation with the comber noil extracted. However, it must be borne in mind that the additional

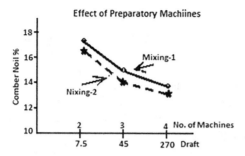

FIGURE 8.16   Preparatory machines and noil%:[2,4] Whenever the number of machines used to prepare the comber lap is increased, the comber extracts lower noil. This is basically due to more parallelization and better orientation.

passages involving a higher total pre-comb draft lead to more parallelization and orientation of fibres in the lap.

Even then, the benefit of saving in the waste owing to additional passages (draft), beyond a certain point, does not bring the same returns. At higher draft levels, the reduction in the comber noil steadily falls down.

8.1.15   Comber Waste[7]

The percentage of waste extracted at the comber mainly depends upon the characteristics of cotton fibre and the extent to which an improvement in the yarn quality is desired. The effect of combing different Indian cottons was studied by Nazir Ahmed[7] with the following findings:

- Extraction of large percentages of comber waste improved the mean length progressively. However, this mean length did not exceed that of raw cotton.
- The fibre length irregularity in the combed material decreased with an increase in comber waste.
- Up to 15% comber waste, there was an improvement in the yarn strength by 10% over that of carded yarns. For still higher waste from 18% to 25%, the strength improvement over that of carded yarns was about 18%. This suggests that improvement in the yarn strength, with the former being due to better fibre orientation and parallelization; whereas with the latter, it was due to improvement in mean length and fibre length regularity.

Further, there was a marked improvement in the yarn appearance when waste extraction levels were higher than 18%. There was also a substantial reduction in the neps in the combed material. For fine and superfine cottons, therefore, the improvement in the yarn and the resulting fabric appearance can bring more value to the goods and would fetch more price.

8.1.16 Waste Level and Comber Settings

A higher comber waste through a set of finer comber needling arrangements (both needle count and their spacing) for superior quality mixings helps in mainly reducing neps. However, the feed, in this case, should not be too heavy. Similarly, higher waste through deeper top comb penetration appears to have a significant effect on the imperfections in the yarn. However, due care must be taken to see that the level of top comb penetration does not affect the orderly arrangement of the comber web during piecing and detachment.

An increase in waste% by widening the detachment setting improves the fibre length uniformity. It is quite obvious that this is due to improved fractionation. It is, therefore, essential to observe the level of fibre length uniformity and impurities in the lap fed to the comber. If the former is poor, it is possible to get improved results by selecting an appropriate step gauge to arrive at a particular comber waste percentage. However, if the impurities and/or neps are more in the material fed, then manipulating waste levels with the top comb (penetration or timing) rather than step gauge brings significant improvements.

The setting between the cushion plate (bottom nipper) and cylinder half-lap is found to have a profound influence on combing quality. A closer setting (less than 25/1000" (or 0.635 mm) increases penetration of the cylinder needles and is expected to remove the short fibres and neps more effectively. However, a closer setting is likely to increase combing force and may cause the plucking of fibres from the nipper grip. Under extreme conditions, it may even lead to fibre breakage. In this respect, the studies carried out at BTRA have failed to show any marked improvement in fibre length characteristics, nep removal and lea C.S.P. With fine cottons, the improvements in U% and thick places were not significant, however, with medium cottons, some improvements in these properties were observed.

8.1.17 Innovation in Comber (Double-Sided Comber)[2,4]

It was Saco–Lowell that first conceived the idea of making a double-sided comber and later introduced it. Since then, they have been the sole manufacturers of a double-sided comber (Figure 8.17). Basically, there are some difficulties in manufacturing a comber of this type, especially the driving of nippers and nipper assembly to open and close the nippers during piecing-up operation and when combing by cylinder needles is carried out.

The machine has six heads on either side. The problem of driving nippers and the whole assembly is tackled in a very ingenious way. In the usual fashion, the nippers are controlled from the nipper shaft. This shaft rotates through a small angle on either side (shown by the arrows). The movement is finally conveyed via N to the nippers.

Accordingly, the right-hand nipper (on one side of the comber) and the left-hand nipper (on the other side of the comber) are swung alternately. As can be seen from Figure 8.17, when the nipper on one side opens, the nipper on the other side is closed. In line with this, is the combing by the cylinder. Again it can be seen from the figure that on one side, the main combing is ongoing, whereas, on the opposite, the piecing-up operation is in progress.

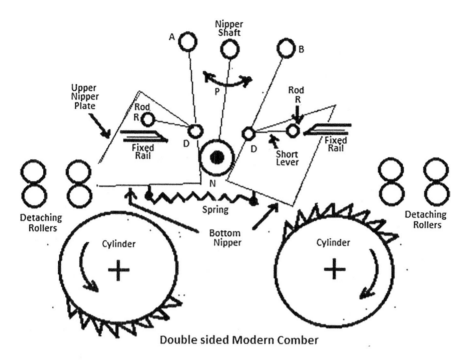

Double sided Modern Comber

FIGURE 8.17   Principle of double-sided comber:[2,4] All traditional and modern combers are one-sided. The double-sided modern comber is expected to give almost double production for a given machine space.

The most innovative feature is the opening and closing of the nippers, which is obtained automatically by the swinging movement (P). When the whole mechanism shifts to the left (with reference to the figure), the small rod (R) rides over a fixed rail. The ultimate linking raises the top nipper (opening of nippers). When the nippers move back (as shown on the right side), the small rod (R) becomes free of that fixed rail and a powerful spring (not shown) pulls down the top nipper on the corresponding bottom nipper. The top combs for the two sides, obviously are separate. The detaching rollers, drafting system and sliver condensation are also separate for the two sides. With two deliveries, the slivers formed are coiled into two separate cans.

REFERENCES

1. *Manual of Cotton Spinning: "Draw Frames, Comber & Speed Frames": Frank Charnley*, The Textile Institute Manchester, Butterworths, 1964
2. *Technology of Short Staple Spinning: W. Klein*, Textile Institute Manual of Textile Technology
3. Rieter Comber E 7/4, Ribbon Lap E 4/1 – Pamphlets, brochures & booklets
4. Elements of Cotton Spinning – Combing - Dr. A.R.Khare, Sai Publication
5. *Spinning: Drawing, Combing & Roving: Book of Papers: Dr. R. Chattopadhyay, Dr. R.S. Rangasamy: NCUTE Programme Series*, 1999
6. Super Lap Former SL 100, Comber VC 5 – Toyota booklet
7. Nazir Ahmed - Tech Bulletin, 1938, Series A, No.44

9 Comber Calculations

9.1 COMBER CALCULATIONS[1,2]

9.1.1 SLIVER LAP MACHINE (GENERAL GEARING)

The gearing of the machine is shown in Figure 9.1. The drive is initiated from the motor and it reaches the machine shaft. The machine shaft drives the lap rollers on one side and the calendar rollers on the other side. The front drafting roller gets its drive from the calendar roller, and in turn, passes it on to the back roller (B.R.), T.R. and S.R. The drive to the lifting rollers is given by the B.R.

9.1.1.1 Speeds

1) $$\text{Machine Shaft} = \frac{950 \times 21}{80} = 249.37 \text{ rpm}$$

2) $$\text{Lap Roller} = \frac{249.37 \times 12}{72} = 41.56 \text{ rpm}$$

3) $$\text{Calender Roller} = \frac{249.37 \times 29}{72} = 100.44 \text{ rpm}$$

4) $$\text{Front Roller} = \frac{100.44 \times 46 \times 40}{20 \times 20} = 462.02 \text{ rpm}$$

5) $$\text{Back Roller} = \frac{462.02 \times 20}{30} = 308.01 \text{ rpm}$$

6) $$\text{Third Roller} = \frac{308.01 \times 26}{24} = 333.67 \text{ rpm}$$

7) $$\text{Sec ond Roller} = \frac{333.67 \times 26}{24} = 361.48 \text{ rpm}$$

8) $$\text{Lifter Roller} = \frac{308.01 \times 32 \times 35}{48 \times 35} = 205.34 \text{ rpm}$$

DOI: 10.1201/9780429486555-9

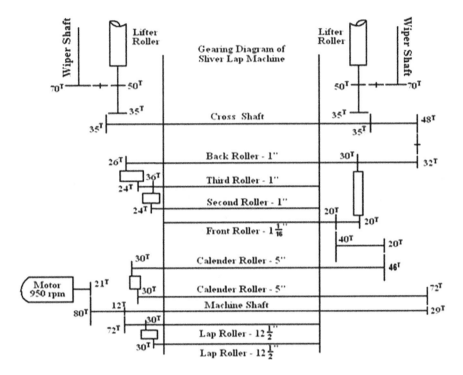

FIGURE 9.1 Sliver lap machine:[1,2] The gearing diagram of a conventional sliver lap machine reveals how the drives are reached to all important parts. It helps in calculating the speeds of these organs.

9.1.1.2 Surface Speeds

1) $$\text{Lap Roller} = \frac{41.56 \times \pi \times 12}{12} \times 0.3048 = 41.45 \text{ m/min}$$

[Note: 1 ft/min = 0.3048 m/min]

2) $$\text{Calender Roller} = \frac{100.44 \times \pi \times 5}{12} \times 0.3048 = 40.07 \text{ m/min}$$

3) $$\text{Front Roller} = \frac{462.02 \times \pi \times 1 - 1/16}{12} \times 0.3048 = 39.17 \text{ m/min}$$

4) $$\text{Back Roller} = \frac{308.01 \times \pi \times 1}{12} \times 0.3048 = 24.57 \text{ m/min}$$

5) $$\text{Third Roller} = \frac{333.67 \times \pi \times 1}{12} \times 0.3048 = 26.62 \text{ m/min}$$

6) $$\text{Second Roller} = \frac{361.48 \times \pi \times 1}{12} \times 0.3048 = 28.84 \text{ m / min}$$

7) $$\text{Lifter Roller} = \frac{204.34 \times \pi \times 1 - 3/8}{12} \times 0.3048 = 22.42 \text{ m / min}$$

9.1.1.3 Draft Between

1) Lap Roller and Calender Roller = 41.45 / 40.07 = 1.03
2) Calender Roller and Front Roller = 40.07 / 39.17 = 1.02
3) Front Roller and Second Roller = 39.17 / 28.84 = 1.35
4) Second Roller and Third Roller = 28.84 / 26.62 = 1.08
5) Third roller and Back Roller = 26.62 / 24.57 = 1.08
6) Back Roller and Lifter Roller = 24.57 / 22.42 = 1.09
7) Total Draft = 1.03 × 1.02 × 1.35 × 1.08 × 1.08 × 1.09
   = 1.80

This is also equal to – (41.45 / 22.42) = 1.8

9.1.1.4 Production Rates

With a lap weight of 35 g/m (approximately 540 grains/yd), the production rate of the machine per shift of 8 hours would be:

$$= \frac{41.45 \times 60 \times 8 \times 35}{1000} = 696.36 \text{ kg / shift}$$

The production thus calculated is for 100% machine efficiency. There are always efficiency losses, owing to lap changes, breakdowns, etc. These may vary depending on the situation existing in the mill. The actual production, therefore would be the calculated production multiplied by the efficiency

The total draft in the machine is usually very small. Whenever it is desired to vary this draft, it is the change pinion (C.P.) (see gearing) which is changed. This changes the draft in the draw box between the front roller and back roller.

Thus, the **Draft Constant** = Total Draft × Change Pinion

$$= 1.8 \times 30 = 54$$

[Note: Here the C.P. is a 30^T wheel on the back roller and it is a driven wheel]

9.1.2   RIBBON LAP MACHINE

9.1.2.1   General Gearing

The drive from the motor pulley is given to the loose and fast pulleys on the machine shaft. This shaft drives the front roller, which in turn, drives a set of drafting rollers. The drive to the creel roller holding the feed laps is given from the back roller through a chain drive. (Figure 9.2)

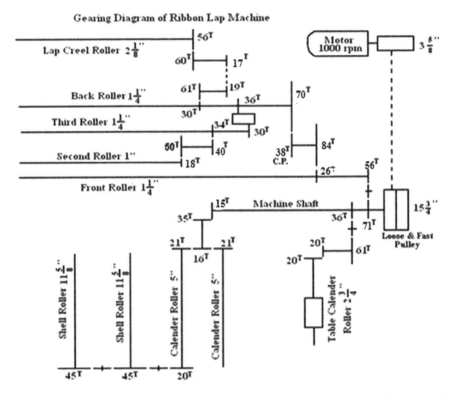

FIGURE 9.2 Ribbon lap machine:[1,2] In the ribbon lap several laps are combined and drafted again to give a more uniform lap of original size and thickness.

The machine shaft also drives the table calender roller and weighted C.R. The drive to shell rollers (on which the ribbon lap is formed) is given from weighted calender rollers.

9.1.2.2   Speeds

1)
$$\text{Machine Shaft} = \frac{1000 \times 3.625}{15.75} = 230.15 \text{ r.p.m.}$$

2)
$$\text{Shell Rollers} = \frac{230.15 \times 15 \times 16 \times 20}{35 \times 21 \times 45} = 33.40 \text{ r.p.m.}$$

3)
$$\text{Weighted C.R.} = \frac{230.15 \times 15 \times 16}{35 \times 21} = 75.15 \text{ r.p.m.}$$

4)
$$\text{Table C.R.} = \frac{230.15 \times 36 \times 20}{61 \times 20} = 135.82 \text{ r.p.m.}$$

5) $$\text{Front Roller} = \frac{230.15 \times 71}{56} = 291.79 \text{ r.p.m.}$$

6) $$\text{Back Roller} = \frac{291.79 \times 26 \times 38}{84 \times 70} = 49.02 \text{ r.p.m.}$$

7) $$\text{Third Roller} = \frac{49.02 \times 36}{30} = 58.83 \text{ r.p.m.}$$

8) $$\text{Second Roller} = \frac{58.83 \times 34 \times 50}{40 \times 18} = 138.90 \text{ r.p.m.}$$

9) $$\text{Lap Roller} = \frac{49.02 \times 30 \times 19 \times 60}{61 \times 17 \times 56} = 28.86 \text{ r.p.m.}$$

9.1.2.3 Surface Speeds

1) $$\text{Shell Roller} = \frac{33.4 \times \pi \times 93}{8 \times 12} \times 0.3048 = 30.98 \text{ m / min}$$

[Note: 1 ft/min = 0.3048 m/min]

2) $$\text{Weighted C.R.} = \frac{75.15 \times \pi \times 5}{12} \times 0.3048 = 29.98 \text{ m / min}$$

3) $$\text{Table C.R.} = \frac{135.82 \times \pi \times 11}{4 \times 12} \times 0.3048 = 29.80 \text{ m / min}$$

4) $$\text{Front Roller} = \frac{291.79 \times \pi \times 5}{4 \times 12} \times 0.3048 = 29.10 \text{ m / min}$$

5) $$\text{Second Roller} = \frac{!38.90 \times \pi \times 1}{12} \times 0.3048 = 11.08 \text{ m / min}$$

6) $$\text{Third Roller} = \frac{58.83 \times \pi \times 5}{4 \times 12} \times 0.3048 = 5.86 \text{ m / min}$$

7) $$\text{Back Roller} = \frac{49.02 \times \pi \times 5}{4 \times 12} \times 0.3048 = 4.89 \text{ m / min}$$

8) $$\text{Lap Roller} = \frac{28.86 \times \pi \times 17}{8 \times 12} \times 0.3048 = 4.89 \text{ m / min}$$

9.1.2.4   Drafts Between
1) Shell Roller and Weighted C.R. = 30.98 / 29.98 = 1.03
2) Weighted C.R. and Table C.R. = 29.98 / 29.80 = 1.006
3) Table C.R. and F.R. = 29.80 / 29.10 = 1.02
4) F.R. and S.R. = 29.10 / 11.08 = 2.62
5) S.R. and T.R. = 11.08 / 5.86 = 1.89
6) T.R. and B.R. = 5.86 / 4.89 = 1.19
7) B.R. and Creeper Roller = 4.89 / 4.89 = 1.0
8) Draw-Box Draft = 29.10 / 4.89 = 5.95
9) Total Draft = 30.98 / 4.89 = 6.33
10) Draft Constant = Total Draft × C. P.
   = 6.33 × 38 = 240.54

For varying drafts, the following equation may be used when the value of the change pinion (C.P.) is required to be changed. This is done for obtaining the desired weight per unit length of the lap.

$$\text{Draft} = \text{(Draft Constant)} / \text{(Change Pinion)}$$

[**Note:** The value of the draft to be put also depends upon the hank of the material fed]

9.1.2.5   Production
The production of the machine for a lap weight of 500 grains/yd or 35 g/m is as follows:

$$\text{Production} = \frac{30.98 \times 60 \times 8 \times 35}{1000} = 520.46 \text{ kg / shift}$$

$$\left(\text{At } 100\% \text{ efficiency}\right)$$

Here again, the production is affected due to the losses in efficiency. These are usually owing to machine downtime for maintenance and repairs, changes in the mixing and package changing time for the feed and delivered laps. The actual production time, therefore, will be less by about 8–10%.

9.1.3   Super Lap Machine[3]

9.1.3.1   General Gearing
The drive from the motor is given to the machine shaft (Figure 9.3) which passes it to the lap roller through the Jack Shaft. The end wheel on the Jack Shaft gives drive to the Stack Calender Rollers. The Miter Box on the short shaft drives the Sliver Table Shaft. There are three heads and each has its own set of Table Calender Rollers and Drafting Rollers.

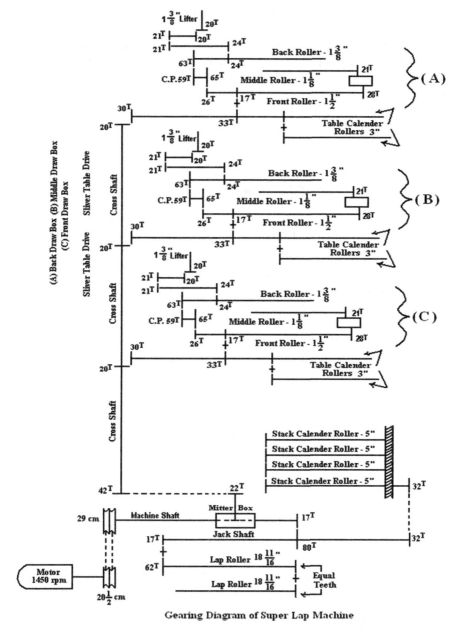

FIGURE 9.3   Super lap machine:[1,3] Whitin introduced their super lap machine to match the performance of the high-speed Whitin comber. The lap regularity and the lap were both improved.

The drive from the Sliver Table Shaft to the Calender Roller pair, in each head, is carried through bevel wheels. The Table Calender Roller drives the Front Roller, which in turn, drives the Back Roller from one end and the second roller from the other end. The drive to the lifter roller is reached from the back roller.

9.1.3.2 Speeds

1) $$\text{Machine Shaft} = \frac{1450 \times 20.5}{29} = 1025 \text{ r.p.m.}$$

2) $$\text{Lap Roller} = \frac{1025 \times 17 \times 17}{80 \times 62} = 59.72 \text{ r.p.m.}$$

3) $$\text{Stack Calender Roller} = \frac{1025 \times 17 \times 32}{80 \times 32} = 217.81 \text{ r.p.m.}$$

4) $$\text{Table C.R.} = \frac{1025 \times \left(\text{Miter Box} = 1\right) \times 22 \times 20}{42 \times 30} = 357.93 \text{ r.p.m.}$$

5) $$\text{Front Roller} = \frac{357.9 \times 33}{17} = 694.8 \text{ r.p.m.}$$

6) $$\text{Middle Roller} = \frac{694.8 \times 28}{21} = 926.4 \text{ r.p.m.}$$

7) $$\text{Back Roller} = \frac{694.8 \times 26 \times 59}{65 \times 63} = 260.27 \text{ r.p.m.}$$

8) $$\text{Lifter Roller} = \frac{260.27 \times 24 \times 21 \times 20}{24 \times 21 \times 20} = 260.27 \text{ r.p.m.}$$

9.1.3.3 Surface Speeds

1) $$\text{Lap Roller or Driving Drum} = \frac{59.72 \times 299 \times \pi}{12 \times 16} \times 0.3048 = 89.06 \text{ m / min}$$

[Note: 1 foot – 0.3048 m]

2) $$\text{Stack Calender Roller} = \frac{217.81 \times 5 \times \pi}{12} \times 0.3048 = 86.91 \text{ m / min}$$

3) $$\text{Table C.R.} = \frac{357.93 \times 3 \times \pi}{12} \times 0.3048 = 85.69 \text{ m / min}$$

4) $$\text{Front Roller} = \frac{694.80 \times 1.5 \times \pi}{12} \times 0.3048 = 83.17 \text{ m / min}$$

5) $$\text{Middle Roller} = \frac{926.4 \times 9 \times \pi}{8 \times 12} \times 0.3048 = 83.17 \text{ m / min}$$

6) $$\text{Back Roller} = \frac{260.27 \times 11 \times \pi}{8 \times 12} \times 0.3048 = 28.55 \text{ m / min}$$

7) $$\text{Lifter Roller} = \frac{260.27 \times 11 \times \pi}{8 \times 12} \times 0.3048 = 28.55 \text{ m / min}$$

9.1.3.4 Drafts Between

1) $$\text{Lap Roller \& Stack C.R.} = \frac{89.06}{86.91} = 1.02$$

2) $$\text{Stack C.R. \& Table C.R.} = \frac{86.91}{85.69} = 1.01$$

3) $$\text{Table C.R. \& F.R.} = \frac{85.69}{83.17} = 1.03$$

4) $$\text{F.R. \& Middle Roller} = \frac{83.17}{83.17} = 1.00$$

5) $$\text{Middle Roller \& B.R.} = \frac{83.17}{28.55} = 2.91$$

6) $$\text{B.R. \& Lifter Roller} = \frac{28.55}{28.55} = 1.00$$

$$\text{Total Draft} = \frac{89.06}{28.55} = 3.118$$

9.1.3.5 Production

The production with a lap weight of 70 g/m (1080 grains/yd) per shift with 100% efficiency will be:

$$\text{Production Rate} = \frac{89.06 \times 60 \times 8}{1000} \times 70 = 3010 \text{ kg / shift}$$

This production is at 100% efficiency level. Depending upon the performance of the machine, there are efficiency tosses due to machine stoppages for various reasons. Therefore, the actual efficiency of the machine will be always lower. Usually, a satisfactory level of efficiency is between 88 and 90%.(for modern machines it is still higher).

9.1.3.6 Changing of Draft in Individual Heads

As there are three heads in the draw box, it is possible to change the draft in individual heads independently.

$$\text{Total Draft in Draw box} = \frac{63 \times 65 \times 3 \times 8}{59 \times 26 \times 2 \times 11} = 2.91$$

Usually, the change wheel (C.P.) is so chosen as to arrive at an appropriate draft level. In certain situations, along with C.P., another combination wheel is also changed to arrive at a precise value of draft in the draw-box. However, while calculating the draft constant, only C.P. is used. Thus, for the above set of wheels:

Draft Constant = Total Draft × C.P. = 3.118 × 59 = 183.96

As mentioned earlier, the draft in the head draw-box can be kept in such a way as to have a slightly higher draft level in the middle head and comparatively lower draft levels in the front and back head. Thus, more parallelization would take place in the middle head which is sandwiched between the two outer layers. In this way, it is possible to reduce the lap-licking tendency. However, it is necessary to select the draft levels on these three heads in such a way that the final lap weight remains equal to the nominal lap weight.

9.1.4 COMBER[1,2]

9.1.4.1 General Gearing for Conventional Comber

The motor drives the machine shaft, which in turn, drives the cylinder shaft and brush shaft (Figure 9.4). The end wheel on the cylinder shaft drives the draw-box cross shaft. The wheel with 72^T on this cross shaft, drives the draw-box calender roller; whereas, another wheel with 34^T on the same cross shaft drives the second drafting roller.

The front roller is driven from the draw-box calender roller. The drive to the third roller, back roller and table calender roller is given from the second roller.

The vertical coiler cross shaft is driven from the draw-box cross shaft. The coiler calender roller is driven from this coiler cross shaft. In the same way, coiler plate and can plate get their drive from vertical coiler shafts.

9.1.4.2 Speeds

(1) $$\text{Machine Shaft} = \frac{960 \times 50}{136} = 352.94 \text{ r.p.m.}$$

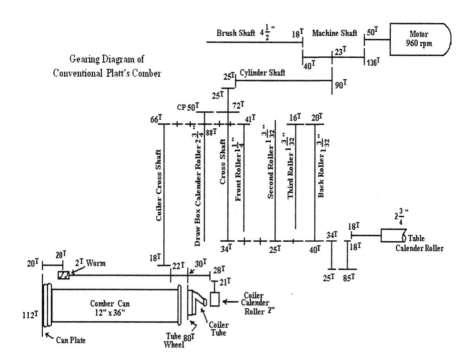

FIGURE 9.4 Gearing diagram of the conventional Platt's comber:[1,2] The most important thing in the comber is the speed of the cylinder which indicates the productive capability of the machine.

(2)
$$\text{Cylinder Shaft} = \frac{352.94 \times 23}{90} = 90.19 \text{ r.p.m.}$$

(3)
$$\text{Brush Shaft} = \frac{352.94 \times 40}{18} = 784.31 \text{ r.p.m.}$$

(4)
$$\text{Draw} - \text{Box C.R.} = \frac{90.19 \times 25 \times 72}{25 \times 50} = 129.87 \text{ r.p.m.}$$

(5)
$$\text{Coiler C.R.} = \frac{129.87 \times 88 \times 18 \times 28}{66 \times 22 \times 21} = 188.90 \text{ r.p.m.}$$

(6)
$$\text{Front Roller} = \frac{90.19 \times 25 \times 72 \times 88}{25 \times 50 \times 41} = 278.75 \text{ r.p.m.}$$

(7)
$$\text{Second Roller} = \frac{90.19 \times 25 \times 34}{25 \times 25} = 122.65 \text{ r.p.m.}$$

(8) $$\text{Third Roller} = \frac{90.19 \times 25 \times 34 \times 20}{25 \times 40 \times 16} = 95.82 \text{ r.p.m.}$$

(9) $$\text{Back Roller} = \frac{90.19 \times 25 \times 34}{25 \times 40} = 76.66 \text{ r.p.m.}$$

(10) $$\text{Table C.R.} = \frac{76.66 \times 40 \times 25 \times 18}{34 \times 85 \times 18} = 26.52 \text{ r.p.m.}$$

(11) $$\text{Coiler Tube Wheel} = \frac{129.87 \times 88 \times 18 \times 30}{66 \times 22 \times 80} = 53.12 \ r.p.m.$$

(12) $$\text{Can Plate} = \frac{129.87 \times 88 \times 18 \times 2 \times 20}{66 \times 22 \times 20 \times 112} = 2.52 \text{ r.p.m.}$$

9.1.4.2.1   Speeds of Various Rollers in One Revolution of Cylinder: This Gives a Fair Idea of the Material Delivered in Each Cylinder Combing Cycle

(1) $$\text{Lap Roller} = \frac{4 \times 48 \times 35}{75 \times 80 \times 46} = 0.0243 \text{ rev.}$$

[Note: The pawl pushes 4^T of the lap roller ratchet which has 75 teeth. The gearing from ratchet to lap roller is as given above in 8.4.2.1 (1)

Similarly, 5^T of the ratchet (47^T) on the feed roller are pulled by the feed roller pawl]

(2) $$\text{Feed Roller} = \frac{5}{47} = 0.085 \text{ rev.}$$

(3) $$\text{Detaching Roll} - (a)(Forward) = \frac{17 \times 50}{30 \times 16} = 1.77 \text{ rev.}$$
$$(b)\text{Backward} = \frac{9 \times 50}{30 \times 16} = 0.93 \text{ rev.}$$

Hence, for a cylinder speed of 90.19 r.p.m. the speeds can be readily calculated:

(1) Lap Roller $= 0.0243 \times 90.19 = 2.19$ r.p.m.

(2) Feed Roller $= 0.085 \times 90.19 = 7.67$ r.p.m.

(3) Detaching Roller $-($Forward Direction$) = 1.77 \times 90.19 = 159.72$ r.p.m.

$\qquad\qquad\qquad -($Backward Direction$) = 0.93 \times 90.19 = 84.50$ r.p.m.

$\qquad\qquad\qquad -($Net Forward Movement$) = 159.72 - 84.50 = 75.21$ r.p.m.

9.1.4.3   Surface Speeds

(1) $\text{Coiler C.R.} = \dfrac{188.90 \times \pi \times 2}{12} = 98.92 \text{ ft/min} = 30 \text{ m/min}$

[Note: 1 foot = 0.3048 m]

(2) $\text{Draw-Box C.R.} = \dfrac{129.87 \times \pi \times 11}{4 \times 12} = 93.51 \text{ ft/min} = 28.36 \text{ m/min}$

(3) $\text{Front Roller} = \dfrac{278.75 \times \pi \times 5}{4 \times 12} = 91.23 \text{ ft/min} = 27.67 \text{ m/min}$

(4) $\text{Second Roller} = \dfrac{122.65 \times \pi \times 35}{32 \times 12} = 35.12 \text{ ft/min} = 10.65 \text{ m/min}$

(5) $\text{Third Roller} = \dfrac{95.82 \times \pi \times 35}{32 \times 12} = 27.44 \text{ ft/min} = 8.32 \text{ m/min}$

(6) $\text{Back Roller} = \dfrac{76.66 \times \pi \times 35}{32 \times 12} = 21.95 \text{ ft/min} = 6.65 \text{ m/min}$

(7) $\text{Table C.R.} = \dfrac{26.52 \times \pi \times 11}{4 \times 12} = 19.09 \text{ ft/min} = 5.81 \text{ m/min}$

(8) Net Forward Movement D.R. $= \dfrac{75.21 \times \pi \times 7}{8 \times 12} = 17.23 \text{ ft/min} = 5.25 \text{ m/min}$

(9) $\text{Feed Roller} = \dfrac{7.675 \times \pi \times 7}{8 \times 12} = 1.758 \text{ ft/min} = 0.53 \text{ m/min}$

(10) $\text{Lap Roller} = \dfrac{2.191 \times \pi \times 11}{4 \times 12} = 1.577 \text{ ft/min} = 0.47 \text{ m/min}$

9.1.4.4   Draft Between

(1) $\text{Coiler C.R. \& Draw-Box C.R.} = \dfrac{30.0}{28.36} = 1.057$

(2) $\text{Draw-Box C.R. \& F.R.} = \dfrac{28.36}{27.67} = 1.024$

(3) $$\text{F.R. \& S.R.} = \frac{27.67}{10.85} = 2.59$$

(4) $$\text{S.R. \& T.R.} = \frac{10.65}{8.32} = 1.27$$

(5) $$\text{T.R. \& B.R.} = \frac{8.32}{6.65} = 1.25$$

(6) $$\text{F.R. \& B.R.}\left(\text{Draw} - \text{Box Draft}\right) = \frac{27.67}{6.65} = 4.15$$

(7) $$\text{B.R. \& Table C.R.} = \frac{6.65}{5.81} = 1.14$$

(8) $$\text{Table C.R. \& Detaching Roller} = \frac{5.81}{5.25} = 1.107$$

(9) $$\text{Detaching Roller \& Feed Roller} = \frac{5.25}{0.53} = 9.9$$

(10) $$\text{Feed Roller \& Lap Roller} = \frac{0.53}{0.47} = 1.11$$

(11) $$\text{Total Draft} = \frac{30.15}{0.48} = 62.81$$

(12) $$\text{Draft Constant} = \text{Total Draft} \times \text{C.P.}\left(50^{\text{T}}\right)$$

Therefore the draft constant = 62.81 × 50 = 3140.5

The total draft of 62.81 is the speed ratio between the coiler calender roller and lap roller, i.e., the final delivery point and starting feeding point. This draft is called a 'mechanical draft' and is decided by the gear wheels of the machine. The machine, however, extracts a certain percentage of waste or noil. Hence, apart from the attenuation by the gears, the thinning-out of the material also occurs owing to the extraction of noil. Thus, the material actually experiences more reduction in thickness than carried out by the mechanical (draft) wheels. A simple relation exists between the 'actual draft', 'mechanical draft' and the noil%.

$$\text{Actual Draft} = \frac{\text{Mechanical Draft}}{\left(100 - \text{Waste\%}\right)} \times 100$$

Thus, with 10% comber noil extracted and the present value of mechanical draft of 62.81, the actual draft will be:

$$\text{Actual Draft} = \frac{62.81}{(100-10)\times100} = 69.78$$

9.1.4.5 Production

The machine production can be calculated from two different sources: (1) from the delivery rate and (2) feed side. With the latter, there is an additional factor required to be considered – noil%.

(1) From delivery rate:

$$\text{Prod. / shift of 8 hrs.} = \frac{\text{Surface speed of C.R.}\times60\times8}{3\times840\times\text{Hank of Silver}} = \frac{98.92\times60\times8}{3\times840\times\text{Hank of Silver}}$$

$$= \frac{18.84}{\text{Hank of Silver}} = \frac{18.84}{0.15\times2.205} = 56.96 \text{ kg} \quad [A]$$

The production is at 100% efficiency assuming the sliver hank as 0.15. Where the gearing is the same for all the machines in a combing department, the constant 18.84 (in production rate [A] above) is useful. This is, therefore called the 'production constant'. When the weight of the sliver is known, the above equation can be modified as:

$$\text{Production / shift} = \frac{\text{Delivery Rate}\left(\text{m / min}\right)\times60\times8}{1000}\times g / m \quad \text{kg / shift}\left(\text{at 100\% Eff}\right)$$

(2) From feed rate:

- Feed/nip (in cm) × nips/min = Feed in cm per min.

Hence,

$$\text{Lap Feed per shift} = \frac{\text{Feed in cm / min}\times60\times8}{100} = \text{metres per shift}$$

- Metres per shift × No. of Heads / machine = Total metres fed / shift / machine

$$\frac{\text{Total metres fed}\times\text{Lap Weight in g / m}}{1000} = \text{Total Weight fed in kg / shift}$$

$$\frac{\text{Kg / shift fed} \times (100 - \text{waste}\%)}{100} = \text{kg per shift delivered}$$

Taking the values from the above examples and from gearing diagram:

Feed per nip = (0.085 × 3.142 × 7/8) = 0.23"/nip = 0.584 cm/nip

With cylinder speed = 90.19 r.p.m; No. of heads = 6; Lap weight = 35 g/m (490 grains/yd and noil% = 10%

- Feed per min per head = (0.584 × 90.19) / (100) = 0.526 m
- Feed/shift/head = 0.526 × 60 × 8 = 252.82 m
- Prod./shift/head = (252.82 × 35) / (1000) = 8.84 kg
- Prod./shift/machine = 8.84 × 6 = 53.04 kg

This is the material fed to the comber. Taking into consideration the noil extracted –

Production (as delivered) = [53.75 × (100 – 10)] / (100) = 48.38 kg – [B]

The production rates in [A] and [B] are different because of different data of hank of sliver and lap weight.

9.1.5   WHITIN COMBER[4]

9.1.5.1   General Gearing

The drive to the cylinder shaft is received from the motor by the 'V' belt pulley. The brush shaft is driven from the cylinder shaft. The bevel at the other end of the cylinder shaft, through a set of gears, drives the fourth drafting roller, which in turn, drives the third and back drafting rollers. Another gear – 57^T – drives the cross shaft, which then drives the tube wheel and coiler C.R. The gearing diagram below shows the 'bi-coiling system'.

The drive to the front roller is received from the other end of cross shaft through $31^T/66^T$ gears. The front roller then drives the second drafting roller. Also, another pair of bevels – 18^T on the cross shaft and 31^T on the short shaft drive can plate through the worm and worm wheel.

There are two separate coilers and the cans to collect two slivers. The hank of the sliver, in this case, is half of the normal sliver weight.

It is thus possible to get the sliver hank in the range of 0.12 to 0.18. It is equally possible to get a much coarser sliver by doubling or combining two slivers (twin coiling) to reduce the number of cans at the creel for post-comb drawing. In such cases, however, there is only one coiler.

9.1.5.2   Speeds

(1) $\text{Machine Pulley} = \dfrac{1440 \times 6.78}{9.90} = 986.18 \text{ r.p.m.}$

(2) $\text{Fan Speed} = \dfrac{1440 \times 6.78}{5.53} = 1746.46 \text{ r.p.m.}$

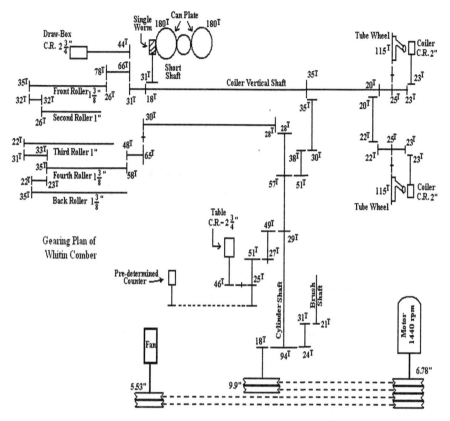

Detaching Roller Diameter = 3.175 cm
Forward delivered length = 5.859 cm
Backward delivered length = 2.336 cm
Net Forward Delivery = 3.522 cm/cycle

FIGURE 9.5   Gearing diagram of a Whitin comber:[1,4] It was the first comber introduced as a then high-speed modern comber. It offered about 50–60% higher productive speeds as compared to the then-conventional comber.

(3)
$$\text{Cylinder} = \frac{986.18 \times 18}{94} = 188.84 \text{ r.p.m.}$$

(4)
$$\text{Brush Shaft} = \frac{188.84 \times 94 \times 31}{24 \times 21} = 1091.82 \text{ r.p.m.}$$

(5)
$$\text{Table C.R.} = \frac{188.84 \times 29 \times 27 \times 25}{49 \times 51 \times 46} = 32.15 \text{ r.p.m.}$$

(6) $\text{Fourth Drafting Roller} = \dfrac{188.84 \times 28 \times 30 \times 48}{28 \times 65 \times 58} = 72.13 \text{ r.p.m.}$

(7) $\text{Back Roller} = \dfrac{72.13 \times 35 \times 22}{23 \times 35} = 68.99 \text{ r.p.m.}$

(8) $\text{Third Roller} = \dfrac{72.13 \times 35 \times 31}{33 \times 22} = 107.79 \text{ r.p.m.}$

(9) $\text{Front Roller} = \dfrac{188.84 \times 57 \times 38 \times 35 \times 31 \times 78}{51 \times 30 \times 35 \times 66 \times 26} = 376.70 \text{ r.p.m.}$

(10) $\text{Second Roller} = \dfrac{376.70 \times 35 \times 32}{32 \times 26} = 507.09 \text{ r.p.m.}$

(11) $\text{Draw} - \text{Box C.R.} = \dfrac{376.70 \times 26 \times 66}{78 \times 44} = 188.35 \text{ r.p.m.}$

(12) $\text{Coiler C.R.} = \dfrac{188.84 \times 57 \times 38 \times 35 \times 23}{51 \times 30 \times 35 \times 23} = 267.33 \text{ r.p.m.}$

(13) $\text{Coiler Tube Wheel} = \dfrac{267.33 \times 23 \times 25}{23 \times 115} = 58.11 \text{ r.p.m.}$

(14) $\text{Can Plate} = \dfrac{58.11 \times 115 \times 18 \times 1}{25 \times 31 \times 180} = 0.86 \text{ r.p.m.}$

The feed roller has a 1" (2.54 cm) diameter and carries a ratchet of 15^T. In every cycle, one tooth of the ratchet is pushed. Similarly, the lap roller has 48^T and has a 2.75" (6.98 cm) diameter. In every cycle, it is pushed for one tooth.

(15) $\text{Feed Roller} = \dfrac{1 \times 188.84}{15} = 12.58 \text{ r.p.m.} = 1.003 \text{ m / min} \left[0.532 \text{ cm / nip} \right]$

(16) $\text{Lap Roller} = \dfrac{1 \times 188.84}{48} = 3.93 \text{ r.p.m.} = 0.862 \text{ m / min}$

9.1.5.3 Surface Speeds

(1) $\text{Coiler C.R.} = \dfrac{267.33 \times 2 \times \pi}{12} \times 0.3048 = 42.67 \text{ m / min}$

[Note: 1 foot = 0.3048 m]

(2) $\text{Draw} - \text{Box C.R.} = \dfrac{188.35 \times 11 \times \pi}{4 \times 12} \times 0.3048 = 41.33 \text{ m / min}$

(3) $\text{F.R.} = \dfrac{376.70 \times 11 \times \pi}{8 \times 12} \times 0.3048 = 41.33 \text{ m / min}$

(4) $\text{S.R.} = \dfrac{507.09 \times 1 \times \pi}{12} \times 0.3048 = 40.46 \text{ m / min}$

(5) $\text{T.R.} = \dfrac{107.99 \times 1 \times \pi}{12} \times 0.3048 = 8.61 \text{ m / min}$

(6) $\text{Fourth Roller} = \dfrac{72.12 \times 11 \times \pi}{8 \times 12} \times 0.3048 = 7.91 \text{ m / min}$

(7) $\text{Back Roller} = \dfrac{68.99 \times 11 \times \pi}{8 \times 12} \times 0.3048 = 7.56 \text{ m / min}$

(8) $\text{Table C.R.} = \dfrac{32.15 \times 11 \times \pi}{4 \times 12} \times 0.3048 = 7.05 \text{ m / min}$

(9) $\text{Feed Roller} = \dfrac{12.58 \times 1 \times \pi}{12} \times 0.3048 = 1.002 \text{ m / min}$

(10) $\text{Lap Roller} = \dfrac{3.93 \times 11 \times \pi}{4 \times 12} \times 0.3048 = 0.86 \text{ m / min}$

(11) $\text{Detaching Roller} = \dfrac{3.522 \times 188.84}{100} \, 6.65 \text{ m / min}$

9.1.5.4 Drafts Between

(1) $\text{Coiler C.R. \& Draw Box C.R.} = \dfrac{42.66}{41.33} = 1.03$

(2) $\text{Draw} - \text{Box C.R. \& F.R.} = \dfrac{41.33}{41.33} = 1.0$

(3)
$$\text{F.R. \& S.R.} = \frac{41.33}{40.46} = 1.02$$

(4)
$$\text{S.R. \& T.R.} = \frac{40.46}{8.61} = 4.70$$

(5)
$$\text{T.T. \& Fourth Roller} = \frac{8.61}{7.91} = 1.08$$

(6)
$$\text{Fourth Roller \& B.R.} = \frac{7.91}{7.56} = 1.04$$

(7)
$$\text{B.R. \& Table C.R.} = \frac{7.56}{7.05} = 1.07$$

(8)
$$\text{Table C.R. \& Detaching Roller} = \frac{7.05}{6.65} = 1.06$$

(9)
$$\text{Detaching Roller \& Feed Roller} = \frac{6.65}{1.003} = 6.63$$

(10)
$$\text{Feed Roller \& Lap Roller} = \frac{1.003}{0.862} = 1.16$$

(11)
$$\text{Total Draft} = \frac{42.66}{0.862} = 49.49$$

The lap fed to this high-speed comber is considerably heavy as compared to the conventional Platt's comber. For example, if a lap of 71 g/m (1000 grains/yd) is fed to the high-speed comber and the machine extracts 12% noil, then the material delivered from each comber head will be:

$$\frac{71 \times (100 - 12)}{100} = 62.48 \text{ g/m}$$

With eight heads per machine, the total material delivered will be:
$$62.48 \times 8 = 499.84 \text{ g/m}$$

Hence, with 49.49 of a draft (mechanical) given in the machine, the material delivered through coiler, as the sliver will be:
$$(499.84) / (49.46) = 10.09 \text{ g/m (142.36 grains/yd)}$$

Considering that this weight/yard of 142.36 grains should represent the sliver, the material passing through the coiler in the form of a sliver would have 0.0585 hank.

[Note: Hank of sliver = 8.33 / grains/yd]

Its weight would be 10.09 g/m OR 10.09 k.tex (Note: k.tex = kg/1000 m)

It should be noted that many modern combers (including the Whitin comber) have a typical bi-coiling arrangement. Therefore, when the same material gets divided into two cans, each with a weight of 5.045 g/m OR 71.165 grains/yd, the hank of each sliver will be 0.117 hank, corresponding to 5.045 k.tex.

9.1.5.5 Production of Comber

(1) **From Coiler C.R.** $= \dfrac{42.67 \times 60 \times 8}{1000} \times 5.045 = 103.33$ kg / shift

As there are two coilers, the total production for the machine will be doubled and will be:

206.66 kg per machine/shift.

  1. **From feed roller:**
  - Feed/nip = 0.532 cm/nip (OR = 0.209"/nip)
  - Nips/min = 188.84
  - No. of head per machine = 8
  - Noil% = 12
  - Lap Weight = 71 g/m (1000 grains/yd)

Production/shift/machine:

$$= \frac{\text{Feed / nip} \times \text{Nips / min} \times 60 \times 8 \times \text{heads / machine} \times \text{Lap Weight} \left(100 - \text{Noil\%}\right)}{100 \times 100 \times 1000}$$

$$= \frac{0.532 \times 188.84 \times 60 \times 8 \times 8 \times 71 \times 88}{100 \times 100 \times 1000} = 529.24 \text{ lbs / shift} = 241.03 \text{ kg}$$

The answer in (a) for the same data of machine particulars of production is 206.59 kg/shift; whereas, in (b) it is 241.03 kg/shift. How can the production calculated from these two sources differ?

It may be recalled that the production calculations in (b) are based on the feeding rate of feed roller and not the lap roller. As quoted above the weight of the material fed is actually the lap weight which is at the lap roller and not at the feed roller. Hence, the weight of the material (lap) at the feed roller will be inversely proportional to the draft between the feed roller and the lap roller. From earlier data, this draft is 1.1635. Thus, the lap of weight 71 g/m , with this draft, will become 71 / 1.16 = 61.2069 g/m at the feed roller. Proportionately, therefore, the production, as calculated in (b) will also reduce by this margin. Hence:

Corrected production at (b) = (241.03) / (1.16) = 207.15 kg/shift

With a difference of decimals, it can now be seen that the production as calculated from (a) and (b) are almost equal.

9.1.6 SOLVED EXAMPLES[1]

Example No. 1: A sliver Lap Machine is fed with 18 slivers, each weighing 3.2 g/m. If the total draft in the machine is 1.8, what is the lap weight delivered?

Total Weight fed = 18 × 3.2; Draft in the machine = 1.8

Therefore, Lap Weight Delivered = (18 × 3.2) / 1.8 = 32 g/m

Example No. 2: The draft constant of the sliver lap machine is 120. If there are 20 slivers each weighing 3.5 g/m fed to the machine so as to get a lap weight of 33.7 g/m, what is the value of the change pinion to be put?

$$\text{Total weight fed} = 3.5 \times 20 = 70 \text{ g/m}$$

$$\text{Total Draft} = \left[\text{Total Weight fed}\right] \div \left[\text{Desired Lap Weight}\right]$$

$$= 70 \div 33.7 = 2.077$$

$$\text{Change Pinion Teeth} = \left[\text{Draft Constant}\right] \div \left(\text{Total Draft}\right) = 120 \div 2.077 = 57.77^{\text{T}}$$

$$= \text{Approx.} 57^{\text{T}}$$

Note: It should be remembered that the sliver lap machine usually has a limit of total draft up to 2.0. Though the value of the draft in the above example is 2.11 this does not exceed the limit by too much; it would be advisable in this case to reduce the weight of the sliver fed to the machine. This would reduce the total draft to be employed, and yet keep the desired lap weight within reach. For the sake of convenience, sometimes, in the mill, people just reduce one sliver (i.e., 19 in place of 20). This is not advisable. Instead, while planning the organization, the hank produced at the card, if possible, should be reduced rather than reducing the number of doublings. This is because; reducing the number of doublings affects the lap regularity.

Thus, for example, if the sliver produced at the card is say 3.0 g/m, the above calculations would be as follows:

$$\text{Total weight fed to sliver lap} = 3.0 \times 20 = 70 \text{ g/m}$$

$$\text{Total draft} = [\text{total weight fed}] / [\text{desired lap weight}] = 60 / 33.7 = 1.78$$

$$\text{C.P.} = [\text{draft constant}] / [\text{total draft}] = 120 / 1.78 = 67.41$$

Therefore, the value of C.P. can be 67^{T} or 68^{T}. With this, there will be only a small change in the weight of the lap delivered.

Example No. 3: Calculate the production of the ribbon lap machine where the lap roller runs at 20 r.p.m. and has a 4.76 cm diameter. The weight of the lap delivered is 29 g/m and the machine has six heads.

Assuming the draft between the final shell roller and lap roller, on which the laps fed to the machine are kept, as six, the delivery rate of the shell roller will be – six times that of the lap roller. The production can, therefore be calculated as follows:

$$\text{Production from Shell roller surface speed} = \frac{20 \times 3.142 \times 4.76\,\text{cm} \times 60 \times 8 \times 29}{100 \times 1000} \times 6$$

$$= 249.82 \text{ kg}$$

Example No.4: A sliver lap machine employs 1.9 draft and delivers a lap of 0.0166 hank (Ne). If the number of slivers fed to the machine is 18, find the hank of the sliver fed.

If 'C' is the hank of the sliver fed then the hank of the material fed will be:

$$= C / 18. \text{ Therefore, with 1.9 draft, this will become:}$$

$$= (C \times 1.9) / 18$$

This means that:

$$C (1.9 \times) / 18 = 0.0166$$

So, C = 0.157

Example No. 5: Comber preparatory machines consist of one draw frame and a sliver lap machine. Find the weight of the lap made for Comber with the following particulars:

Hank of card sliver = 0.13 Ne
Number of doublings on draw frame = 8
Draft in draw frame = 8.2
Number of doublings on sliver lap = 20
Draft in sliver lap = 2.0

The hank of the sliver made on the draw frame will be:

$$\frac{\text{Hank of Card sliver} \times \text{Draft in Draw Frame}}{\text{Doubling}} = \frac{0.13 \times 8.2}{8} = 0.133 \text{ Ne}$$

$$\text{Hank of Sliver Lap Doubling} = \frac{\text{Hank of Drawing sliver} \times \text{Draft}}{20} = \frac{0.133 \times 2}{20} = 0.0133$$

Weight of the comber lap = (8.33) / (0.0133) = 626.31 grains/yd
[Note: 8.33 / hank of material = grains/yd]

Example No. 6: In the comber section, there are 20 high-speed combers, each producing 270 kg/shift. If the lap roller of the super lap machine runs at 60 r.p.m. and has a 40 cm diameter to produce laps of 51 g/m, find the number of super lap machines needed to meet the comber production.

Production of 20 high-speed combers = 280 × 20 = 5600 kg/shift

$$\text{Production of Super Lap m / c} = \frac{60 \times 40 \times 3.142 \times 60 \times 8 \times 51}{100 \times 1000} = 1845.98 \text{ kg / shift}$$

$$\text{Number of Super Lap machines required} = (5600) \div (1845.98) = 3.03$$

$$= \text{nearly 3 machines}$$

Many a time, this problem is faced in mills where the situation is somewhat similar to what has been described in the above example. The requirement of the number of super lap machines is a little more than three. Thus, the three machines in shift would be producing a deficit of just 63 kg. Logically, there are three options open:

1. To speed up the super lap approximately by 2.5%
2. To stop one comber per shift
3. To increase the lap weight by about 2.5%

By suitably selecting the motor/machine pulley dimensions, it may be possible to increase production by the required margin. It may be noted here that the increase in speed, in this case, is quite insignificant and may not exceed the limiting speed of the machine recommended by the manufacturer. If it does in any other case, the option of increasing speed will have to be discarded.

Stopping one comber per shift is only a negative solution and is possible only when the mill is not pressed for the production of combed yarn.

It may also be possible to increase the lap weight through the required margin and this perhaps is the easiest and more logical solution. Modern combers are designed to take higher lap weight even beyond 51 g/m and increased weight by 4–5 grams, therefore, this should not pose any problem for the nippers to grip the lap or cylinder needles (UniComb half-lap) to penetrate it.

Useful Formulae and Conversions:

$$7000 \text{ grains} = 1 \text{ lb}; 2.205 \text{ lbs} = 1 \text{ kg}; 15.36 \text{ grains} = 1 \text{ gram}$$

$$8.33 \text{ grains/yd} = \text{hank (Ne)}, \text{ft/min} \times 0.3048 = \text{m/min}$$

9.1.7 EXERCISES

(1) Find the production per shift of a sliver lap machine when its lap roller of 12" diameter runs at 50 r.p.m. to produce a lap of 400 grains/yd with 80% efficiency. If the length of the lap on each spool is 120 yards, find both the weight of the material on each spool and the time taken to complete one spool.
   [Ans: 521.12 kg/shift; 3.1 kg/lap and 2.29 min/spool]
(2) The lap feed roller of the ribbon lap machine has a 2.75" diameter and runs at 25 r.p.m. If the draft constant of the machine is 400 and C.P. with 52^T

is working, find the production rate per shift with 84% efficiency and 520 grains/yd of lap weight. If the draft between the final lap roller and F.R. is 1.5, also find the F.R. r.p.m. with its diameter as 1.25".

[Ans: 626.78 kg/shift and 236.79 r.p.m.] 7.69 – 38.75

(3) In the super lap machine, the lap roller runs at 60 r.p.m. and has a 15.8" diameter. The lifter roller at the back runs at 270 r.p.m. and has a 1.375" diameter. Find the total draft. When 48 slivers, each with a 0.18 hank, are combined, also find the lap weight/yard.

[Ans: 2.55 draft and 870.96 grains/yd]

(4) A sliver lap machine is fed with a sliver of 3.5 g/m. The draft constant of the machine is 80 and C.P. of 44^T is working. If the number of slivers doubled is 22, find the hank of the lap delivered. 1.81, 11.75 m

[Ans: 42.54 g/m and 0.0117 Nf]

(5) The hank of the lap fed to the ribbon lap machine is 50 k.tex, whereas the hank of the lap delivered is 48.8 k.tex. If the machine has six heads, find the total draft in the machine. If the draft constant is 328, also find the C.P. required.

[Ans: 6.147 draft and 53.35^T C.P.]

Note: As the value of C.P. is odd, it will have to be 53^T or 54^T C.P. But with this, the lap hank will slightly change.

(6) A super lap machine has to meet the production of 5700 kg per shift for the comber department. The lap weight required for the comber laps is 1100 grains/yd. A winding drum of the machine has a 15" diameter and runs at 70 r.p.m. Find the production of the super lap and the number of machines required, if its efficiency is 92%

[Ans: 1.97 m/cs and 2884.08 kg/shift]

9.1.8 SOLVED EXAMPLES – COMBER

Example No. 1: Platt's comber works with the following particulars:
Lap weight = 30 g/m; Feed/nip = 0.45 cm; Nips/min = 98; Noil% = 14; Number of heads = 6 and Efficiency = 81%.

Find the production per shift. If the machine has a metric hank meter, find the number of hanks registered at the end of the shift with a machine draft of 50.

$$\text{Production} = \frac{0.45 \times 98 \times 60 \times 8 \times 30 \times 6 \times (100-14) \times 81}{100 \times 1000 \times 100 \times 100} = 26.54 \text{ kg / shift}$$

The lap length fed by one of the machine head will be:

$$\frac{0.45 \times 98 \times 60 \times 8}{100} = 211.68 \text{ m}$$

So, the length in hanks fed at lap roller = $\dfrac{211.68}{1000} = 0.212$ metric hanks

As the machine employs a mechanical draft of 50:

This length of 0.212 metric hanks will be attenuated to = 0.212 × 50 = 10.6 metric hanks

At 81% efficiency, this will be = 10.6 × 0.81 = 8.58 metric hanks

Hence the machine will register 8.58 metric hanks on the hank meter at the end of the shift.

Example No.2: A coiler C.R. having a 2" diameter of comber runs at 160 r.p.m. and delivers a sliver of 55 grains/yd. If there are six heads to the machine employing a mechanical draft of 48, find the lap weight and production per shift. The expected hanks per shift delivered by the comber are 15.96, whereas, the machine actually gives only 13 hanks per shift with 12% noil extraction.

$$\text{The speed of the coiler C.R.} = \frac{3.142 \times 2 \times 160}{36} = 27.93 \text{ yd / min}$$

$$\text{Production / } shift = \frac{27.93 \times 60 \times 8 \times 55}{2.205 \times 7000} = 47.77 \text{ kg / shift}$$

But this production is at 100% efficiency level. The actual efficiency can be calculated as:

Efficiency = (13 × 100) / 15.96 = 81.45%

Hence the actual production will be = (47.77 × 81.45) / 100 = 38.90 kg/shift

Similarly,
$$\text{Actual Draft} = \frac{\text{Mechanical Draft} \times 100}{(100 - \text{waste\%})} = \frac{48 \times 100}{88}$$

Also,
$$-54.54$$

$$\text{Lap Weight fed} = \frac{\text{Wt. of sliver / yard} \times \text{Actual Draft}}{\text{No. of Doubling}} = \frac{55 \times 54.54}{6} = 499.95 \text{ grs / yd}$$

Example No. 3: A feed roller of 1" diameter on a Whitin comber has a 15^T ratchet. The detaching roller of 1.375" diameter rotates 20^T in the forward direction and 9^T in the backward direction. If this gear on a detaching roller has 38^T, find the draft between the detaching roller and the feed roller.

The feed roller rotates 1 tooth in every combing cycle. Hence feed per nip will Be –

Feed / nip = (1 × 3.142) / 15 = 0.209"

$$20^T \text{ forward means} = \frac{20 \times 11 \times 3.142}{8 \times 38} = 2.273 \text{ inch}$$

$$9^T \text{ backward means} = \frac{9 \times 11 \times 3.142}{8 \times 38} = 1.023 \text{ inch}$$

The net forward motion, therefore will be: (2.273) – (1.023) = 1.25" per nip

Hence the draft between detaching roller and feed roller is

= (1.25) / 0.209 = 5.98

Example No. 4: A super lap machine produces 1300 kg/shift. The hank of comber sliver produced is 0.075 hank. If the Comber feed roller feeds 0.22" of lap per revolution of cylinder and mechanical draft in between (a) feed roller and lap roller = 1.05 and (b) coiler C.R. and lap roller = 52; find the number of comber machines required. The comber extracts 13% noil, has 8 heads and runs at 160 nips/min

The total draft between coiler C.R. and lap roller =

Draft between coiler C.R. and feed roller × draft between feed roller and lap roller

i.e., 52 = coiler C.R. and feed roller × 1.05

Therefore, the draft between coiler C.R. and feed roller = 52 / 1.05 = 49.52

This is a mechanical draft. It will be interesting to also find the actual draft with the given noil%. The actual draft will depend upon the noil extracted. Thus:

$$\text{Actual draft} = \frac{\text{Mechanical Draft}}{(100 - \text{noil}\%)} \times 100 = \frac{49.92}{(100 - 13)} \times 100 = 57.37$$

However, while calculating the speed of coiler C.R., it is the mechanical draft which must be taken into account. This is because it is the speed ratio.

Delivery rate at coiler C.R. = [0.22 × 160 × 49.52] / 12 = 145.25 ft/min

Production with this delivery rate of coiler calender roller and with 0.075 hank of sliver will be:

$$\text{Production} = \frac{145.25 \times 60 \times 8}{3 \times 840 \times 0.075 \times 2.205} = 167.29 \text{ kg / shift}$$

$$\text{No. of Comber machines} = \frac{1300}{167.29} = 7.77 \text{ comber machines.}$$

It may be noted that the hank of the sliver in the above example is twice the thickness of a normal comber sliver. In this example, it is assumed that the bi-coiling system is not working and a single sliver is delivered into the can. In the above example, lap weight is not given; instead, the hank of the sliver as delivered is mentioned. Hence the data regarding noil percentage would be redundant while calculating comber production from the speed of the coiler calender roller. However, it would be useful in calculating back the lap weight.

Therefore, with a total mechanical draft of 52 between the coiler C.R. and lap roller, and 13% noil extracted, the actual draft will be:

$$\text{Actual Draft} = \frac{52}{(100 - 13) \times 100} = 59.77$$

The lap weight for 8 laps fed at the back of the comber will be as follow:

With a sliver hank of 0.075, its weight will be (8.33) /0.075 = 111.06 grains/yd

With the above actual draft, the lap weight of 8 laps will be:

59.77 × 111.06 = 6638.05 grains/yd

Weight/yard of each lap = 6638.05 / 8 = 829.80 grains/yd

Another method for solving Example No. 4:

This is based on the particulars from the feed side. For this, it is necessary to calculate the lap weight from the sliver hank, noil extracted and the mechanical draft.

Sliver weight = 8.33 / 0.075 = 111.06 grains/yd

With a mechanical draft of 52, noil% of 13 and above sliver weight and number of doublings is 8, we have:

$$\text{Lap Weight} = \frac{111.06 \times 52 \times 100}{8 \times (100 - 13)} = 829.75 \text{ grains / yard}$$

Thus, the comber production from the feed side will be:

$$= \frac{\text{Feed / nip} \times \text{nips / min} \times 60 \times 8 \times \text{Lap wt.} \times \text{No. of heads} \times (100 - W\%)}{36 \times 7000 \times 1000 \times 2.205}$$

$$= \frac{0.22 \times 160 \times 60 \times 8 \times 829.75 \times 8 \times 87}{36 \times 7000 \times 100 \times 2.205} = 175.60 \text{ kg / shift}$$

It may be again noted that the lap weight calculated here is that of the actual lap put on the wooden lap roller of the comber, whereas, the other particulars are taken from the feed roller. Hence, it will be necessary to incorporate the draft of 1.05 between the lap roller and feed roller to correct this figure of production.

Hence production will be: (175.60) / 1.05 = 167.52 kg/shift

The number of Combers required will again be:

= (1300) / (167.29) = 7.77 comber machines, i.e., nearly eight machines

Example No. 5: The hank meter on the comber registers 22.7 hanks at the end of the shift. The hank of sliver delivered by the machine is 0.16. If the coiler C.R. has a 2" diameter and runs at 240 r.p.m., find the production per shift of 8 hours and the working efficiency of the machine. The comber has a bi-coiling system in operation.

Note: The hank meter reading indicates the actual length delivered in a given time. Therefore, it takes into consideration the machine's working efficiency.

Hence, production per coiler per shift = 22.7 hanks, i.e., 45.4 hanks per machine.

$$\text{Production per machine} = \frac{\text{Total hanks}}{\text{Hank of silver}} = \frac{45.4}{0.16}$$

$$= 171.87 \text{ lb} \quad \text{or} \quad 64.34 \text{ kg / shift}$$

The calculated length (expected) delivered by each coiler (equivalent to a hank meter reading) is:

$$= \frac{3.142 \times \text{dia.} \times \text{r.p.m.} \times 60 \times 8}{36 \times 840} = \frac{3.142 \times 2 \times 240 \times 60 \times 8}{36 \times 840}$$

$$= 23.93 \text{ hanks}$$

$$\text{Therefore the Efficiency} = \frac{22.7}{23.93} \times 100$$

$$= 94.86\% \text{ Expected Hanks}$$

Example No. 6: A lap of 500 grains/yd is fed to a comber with six heads per machine. The draft constant of the machine is 2000 and 40^T C.P. is working on the machine. If the hank of the sliver delivered is 0.154, find the comber noil%.

$$\text{Mechanical draft} = (2000) / (40) = 50$$

$$\text{Material fed to comber} = \text{Lap weight} \times \text{No. of doublings}$$
$$= 500 \times 6 = 3000 \text{ grains/yd}$$

$$\text{The weight of the sliver} = 8.33 / 0.154 = 54.09 \text{ grains/yd}$$

$$\text{Actual Draft} = \frac{\text{Weight of the material fed / length}}{\text{Weight of the sliver / length}} = \frac{500 \times 6}{54.09}$$

-55.46

$$\text{Actual Draft} = \frac{\text{Mechanical Draft}}{(100 - \text{waste}\%)} \times 100$$

$$55.46 = \frac{50}{100 - \text{waste}\%} \times 100$$

Hence waste% = 55.46 × 100 − 55.46 × waste% = 50 × 1000
   Therefore, 5546 − 5000 = waste% × 55.46
   Hence waste% = 546 / 55.46 = 9.84%

Example No. 7: A high-speed comber is fed with eight laps each of 18 kg. The comber works at 285 nips and has a lap weight of 950 grains/yd. The feed per nip is 0.202", whereas, noil extracted is 12%. If the working efficiency is 85%, find production/shift and time taken for complete exhaustion of one full lap.

$$\text{Production/head} = \frac{0.2 \times 285 \times 60 \times 8 \times 950 \times (100 - 12) \times 85}{36 \times 7000 \times 100 \times 100 \times 2.205}$$

$$= 34.98 \text{ kg / shift}$$

Production per machine = 34.98 × 8 = 220.98 kg/shift
   This is the production delivered by the machine, whereas the material fed by the lap roller will be: [production delivered] / [(100 − waste%) / 100]

$$= 220.98 / (100 - 12) / 12 = 279.91 \text{ kg/shift.}$$

Thus,
Total material in the form of laps to be fed = lap weight × No. of heads

$$= 18 \times 8 = 144 \text{ kg.}$$

Hence, the time taken in a shift of 8 hours to completely exhaust the full lap will be:
Time in shift of 8 hours = (144) / (279.91) = 0.514 shifts OR

$$= 4.1 \text{ hours}$$

Example No. 8: A comber has 6 heads and runs at 100 nips per minute. The noil study taken for short and long duration (i.e., 50 nip-working and 1 hour study respectively) gives the following results:
Find the average noil percentage and range between maximum and minimum waste level.
Short study: The total sliver collected = 1000 grains. When apportioned to each head, the average sliver weight for each will be: (1000) / 6 = 166.66 grains.
Long study: Similarly, the average weight per head will be: 20 / 6 = 3.33 kg
The head-wise comber noil% is expressed as:

$$\text{Head noil\%} = \frac{\text{Weight of noil}}{166.66 + \text{weight of noil}} \quad \text{For Short Study}$$

$$= \frac{\text{Weight of noil}}{3330 + \text{weight of noil}} \quad \text{For Long Study}$$

However, when a more reliable estimate of average comber noil or even the range between the heads is required, it is necessary to take a long study for at least 1–2 hours or even for a full lap. Incidentally, in such cases, rather than taking the weight of the sliver, the lap weights are taken at the beginning and at the end of the study (if only part of the full lap is worked). Thus, it would be easy to find the lap material

TABLE 9.1
Comber Noil Study

Head No.	Short Study	Sliver Collected	Long Study	Sliver Collected
1.	18 grs	Total Sliver	350 g	Total Sliver
2.	17 grs	Collected	350 g	Collected
3.	18 grs	1000 grains	330 g	20 kg
4.	17.5 grs		340 g	
5.	18.5 grs		345 g	
6.	17 grs		360	

TABLE 9.2
Comber Noil Study

	Noil For			
	Short Duration		**Long Duration**	
Head No.	Wt. of Noil	Noil%	Wt. of Noil	Noil%
1.	18.0	9.74	350	9.51
2.	17.0	9.25	330	9.01
3.	18.0	9.74	350	9.51
4.	17.5	9.50	340	9.26
5.	18.5	9.99	345	9.38
6.	17.0	9.25	360	9.75

Range = Maximum Noil% - Minimum Noil%
For Short Duration Range = 9.99 – 9.25 = 0.74%
For Long Duration Range = 9.75 – 9.01 = 0.74%
Note: For a quick estimation of noil level, many a time, the comber is run for a few nips (50 nips in this
example). The aspirator drum is first cleaned before starting the test. Both, the noil from each
head and the sliver coming from the respective head are collected separately. The estimation of
the noil as shown in the above table can then be made.

fed during the course of the study. The expression for the noil% for the respective
head will be:

Head-wise noil% = {[Weight of noil collected] / [Weight of lap cotton fed]} × 100

Example No. 9: In a trial for finding out comber noil, the lap weights before and
after the test for the respective heads were as follows:

(1) 10.9 kg and 8.2 kg; (2) 11.2 kg and 8.45 kg; (3) 9.3 kg and 6.85 kg; (4) 5.4 kg and
2.95 kg; (5) 6.8 kg and 4.2 kg and (6) 8.7 kg and 6.15 kg

The noil collected during the test period for the respective heads was – 300 g, 320
g, 280 g, 310 g, 315 g and 295 g. Find the average noil%, individual head noil% and
range.

The average noil% = 11.66% and range = 12.40 – 11.11 = 1.29%

Example No.10: A high-speed comber with eight heads, running at 275 nips/min
extracts 14% noil. The lap fed to the machine weighs 62 g/m The metric hank meter
on the machine registers 29 hanks at the end of a shift of 8 hours; whereas, the sliver
hank is 0.13 Nf. If the working efficiency of the comber is 92%, find the mechanical
draft and feed per nip for this machine.

29 metric hanks = 29 × 1000 m and with bi-coiling, the total length delivered
through two coilers will be equal to 29000 × 2 m/shift

This is at a 92% efficiency level and hence at a 100% level the length delivered
will be:

TABLE 9.3
Comber Noil Study

Head No.	Material Before – kg	After – kg	Lap Fed – kg (a)	Noil – g (b)	Noil% (b)/(a) × 100
1.	10.9	8.20	2.70	300	11.11
2.	11.2	8.45	2.75	320	11.63
3.	9.30	6.80	2.50	280	11.20
4.	5.40	2.90	2.50	310	12.40
5.	6.80	4.20	2.60	315	12.11
6.	9.70	7.15	2.55	295	11.56

58000 / 0.92 = 63043.48 m
0.13 metric hanks means → (0.13 × 1000 m) weigh 500 g or = 0.5 kg

$$\text{Hence the production} = -\frac{63043.48 \times 0.5}{1000 \times 0.13} = 30.30 \text{ kg per head.}$$

= 242.46 kg per machine per shift delivered
With 14% noil extracted by the comber, the material fed to the machine will be:

$$\frac{242.46}{(100-14)} \times 100 \quad = 281.93 \text{ kg per shift OR } 35.24 \text{ kg per head (A)}$$

From the given machine particulars, the material fed to the comber will be:

$$= \frac{\text{Feed/nip} \times \text{nips/ min} \times 60 \times 8 \times g / m}{100 \times 1000}$$

$$= \frac{\text{Feed/nip} \times 275 \times 60 \times 8 \times 62}{100 \times 1000}$$

= Feed/nip × 81.84 which is equal to production calculated earlier as per (A)
Hence, feed per nip = 35.24 / 81.84 = 0.43 cm (0.17")
Now, the sliver weight delivered can be calculated as:

$$\frac{0.5}{0.13} = 3.84 \, g / m$$

- With bi-coiling sliver weight it will be – 2 × 3.84 = 7.68 g/m

[Note: In metric system (Nf), hank = 0.5 / g/m]

$$\text{Actual Draft} = \frac{\text{Lap Weight fed} \times \text{No. of Doubling}}{\text{Sliver Weight delivered}}$$

$$= \frac{62 \times 8}{7.68} = 64.58$$

$$\text{Mechanical Draft} = \text{Actual Draft} \times \frac{(100 - 14)}{100} = 64.58 \times \frac{86}{100} = 55.53$$

Example No. 11: A comber with six heads has a lap fed with 30 g/m. It is necessary to produce a sliver of 3.5 k.tex with noil extracted as 15%. If the draft constant of the machine is 2100, find the value of C.P.

3.5 k.tex means, a sliver weighing 3.5 kg per 1000 m, i.e., 3.5 × 1000 g / 1000 m

$$\text{Actual Draft} = \frac{\text{Weight of Lap fed/Length}}{\text{Weight of Sliver delivered/Length}} \times \text{No. of Doublings}$$

$$= \frac{30 \times 6}{3.5} = 51.428$$

$$\text{Mechanical Draft} = \frac{\text{Actual Draft} \times (100 - \text{Waste}\%)}{100} = \frac{51.428 \times (100 - 15)}{100} = 43.71$$

$$\text{Change Pinion (CP)} = \frac{\text{Draft Constant}}{\text{Mechanical Draft}} = \frac{2100}{43.71} = 48.04 \ i.e. \ 48T$$

Example No. 12: The speed of the coiler calender roller of a comber is 200 yards/min. It is required to produce 180 kg/shift. The machine extracts 13% noil and has six heads. The cylinder speed is 220 r.p.m (nips/min). If the feed/nip is 0.21", find the lap weight and the hank of the sliver delivered.

Production from the coiler calender roller:

$$= \frac{200 \times 60 \times 8}{3 \times 840 \times \text{Hank of Sliver} \times 2.205} = \frac{17.276}{\text{Hank of Sliver}}$$

Equating this with the actual production:

$$\frac{17.276}{\text{Hank of Sliver}} = 180 \text{ Hence, the hank of sliver} = 0.0959$$

[Note: No bi-coiling]
Production from feed side:

$$= \frac{0.21 \times 220 \times 30 \times 8 \times \text{Lap Weight} \left(grs \ / \ yd \right) \times 8 \left(\text{heads} \right) \times \left(100 - \text{Waste}\% \right)}{36 \times 7000 \times 2.205 \times 100}$$

= lap weight in grains/yd × 0.277

Again equating this with the given production of 180 kg/shift:

Lap weight in grains/yd = (180) / (0.277) = 649 grains/yd

Example No. 13: Total mechanical draft in comber is 50. There is a tension draft of 1.05 each between: (1) feed roller and lap roller; (2) table calender roller and detaching roller; (3) back toller and table calender roller; (4) calender roller and front roller and (5) coiler calender roller and calender roller. If the front roller is 1–1/4" runs at 370 r.p.m; whereas the back toller is 1–1/8" runs at 70 r.p.m, find the draft between detaching roller and feed roller.

The total draft as mentioned above from (1) to (5) will be:

$(1.05)^5 = 1.276$ (A)

$$\text{Draft in the Draw Box}=\frac{\text{F.R. dia.}\times\text{its rpm}}{\text{B.R. dia.}\times\text{its rpm}}=\frac{1-\frac{1}{4}\times370}{1-\frac{1}{8}\times70}=5.873 \quad (B)$$

Total draft = Draft between detaching roller and feed roller × (A) × (B)

50 = Draft between detaching roller and feed roller × 1.276 × 5.873

Hence, draft between detaching roller and feed roller = (50) / (1.276 × 5.873)

Thus, the draft between detaching roller and feed roller = 6.672

9.1.9 SOME TYPICAL EXAMPLES OF FRACTIONATING EFFICIENCY

Example No. 1: Simpson's Method

For finding the two indices – 'combing efficiency' and 'detaching efficiency', the data regarding the fibre length distribution of lap and sliver were collected as follows:

$$(1)\quad \text{Combing Efficiency} = \frac{\sum_{\ell_1}^{\ell_m}\frac{a-A}{\ell}}{\sum_{\ell_1}^{\ell_m}\frac{a}{\ell}}\times100 = \frac{\frac{8-6}{10}+\frac{5-3}{6}+\frac{5-2}{2}}{\frac{8}{10}+\frac{5}{6}+\frac{5}{2}}\times100 = 49.19\%$$

$$(2)\qquad \text{Detaching Efficiency} = \frac{\sum_{\ell_m}^{\ell_{max}}A\times\ell}{\sum_{\ell_m}^{\ell_{max}}a\times\ell}\times100$$

Detaching Efficiency

$$=\frac{(6\times34)+(8\times30)+(18\times26)+(16\times22)+(13\times18)+(10\times14)}{(7\times34)+(10\times30)+(20\times26)+(18\times22)+(15\times18)+(12\times14)}\times100 = 86\%$$

TABLE 9.4

Baer Sorter Diagram Particulars for Calculation of Fractionating Efficiency

Fibre Group Length (mm)	Mean Length of Group mm (ℓ)	% fibres in Lap (a)	% of Fibres in		Actual % of Fibres in Combed Sliver (A)
			Ideal Sliver	Ideal Noil	
1) 32–36	34	07	07	–	06
2) 28–32	30	10	10	–	08
3) 24–28	26	20	20	–	18
4) 20–24	22	18	18	–	16
5) 16–20	18	15	15	–	13
6) 12–16	14	12	12	–	10
7) 08–12	10	08	–	08	06
8) 04–08	06	05	–	05	03
9) 00–04	02	05	–	05	02

Example No. 2: Fractionating efficiency index (F.E.I.) – Parthasarathy's method

There are two terms involved here. F(i) is the factor which is a measure of the capacity of the comber to detach longer fibres in combed sliver. F(ii) is the factor connected with the performance of the comber to remove short fibres from the lap into noil. Thus:

$$F(i) = \frac{\sum_{\ell_m}^{\ell_{max}}(\ell_x - \ell_m)Ax}{\sum_{\ell_m}^{\ell_{max}}(\ell_x - \ell_m)ax} \qquad F(ii) = \frac{\sum_{\ell_1}^{\ell_m}(\ell_m - \ell_x)Ax}{\sum_{\ell_1}^{\ell_m}(\ell_m - \ell_x)ax}$$

Here:

Ax = % of fibres of length ℓx

in Combed Sliver

ax = % of fibres of length ℓx

in original lap

ℓ_{max} = Maximum fibre length

ℓ_m = Boundary length below which fibres

should be removed into noil

ℓ_1 = Minimum length of fibres

in lap & Sliver

Thus, composite F.E.I. can be defined as:

$$F.E.I. = [F(i) - F(ii)] \times 100$$

As can be seen, this composite index takes into account both (a) the positive contribution of the comber in detaching all the longer fibres and (b) the negative contribution in taking all the shorter fibres. With the following data, these two indices are worked out [$\ell m = 12$ mm]

$$F(i) = \frac{(22\times6)+(18\times8)=(14\times18)+(10\times16)+(6\times13)+(2\times10)}{(22\times7)+(18\times10)+(14\times20)+(10\times18)+(6\times15)+(2\times12)} = 0.865$$

$$F(ii) = \frac{(2\times6)+(6\times3)+(10\times2)}{(2\times8)+(6\times5)+(10\times5)} = 0.521$$

$$F.E.I. = \left[F(i)-F(ii)\right]\times100$$

$$= \left[0.865-0.521\right]\times100 = 34.4\%$$

9.1.10 EXERCISES

Example No. 1: A comber works with the following particulars:
  (a) Feed per nip = 0.2"
  (b) Nips/min = 120
  (c) Noil% = 11.5

TABLE 9.5
Baer Sorter Diagram Particulars for calculation of Fractionating Efficiency

Fibre Length Group (mm)	Mean Group Length mm (ℓx)	% if Fibres in Lap (ax)	% of Fibres in Sliver (Ax)	Weightage Factor (ℓx − ℓm)
32–36	34	07	06	22
28–32	30	10	08	18
24–28	26	20	18	14
20–24	22	18	16	10
16–20	18	15	13	06
12–16	14	12	10	02
08–12	10	08	06	02
04–08	06	05	03	06
00–04	02	05	02	10

ℓx = Mean Group Length
ℓm = Limiting Boundary Length

(d)　No. of Heads = 6

(e)　Efficiency = 76%

(f)　Lap weight = 560 grains/yd

Find the comber production per hour and expected hanks at the end of the shift of 8 hours, if the mechanical draft in the comber is 45.

　　　[Ans: 5.85 kg/hr; 13.02 hanks/shift]

Example No. 2: A comber with six heads has a draft constant of 1980 and works with 44^T C.P. The machine is fed with 500 grains/yd lap and the sliver delivered is 0.145 hank. Find the percentage noil extracted.

　　　[Ans: 13.83%]

Example No. 3: A feed roller delivers 0.18"/nip. The 45^T wheel on detaching roller turns 11^T in the backward direction and 25^T in the forward direction. If detaching roller has a 1.0625" diameter, find the draft between the detaching roller and feed roller. Also, find the net delivered length by the detaching roller in each combing cycle.

　　　[Ans: Draft = 5.77; Net length delivered =1.038"/nip]

Example No. 4: The laps, each weighing 10 kg are fed to a comber having eight heads and running at 150 nips/min. The lap roller has a 2.5" diameter and carries a ratchet of 75 teeth. A pawl on this ratchet pushes 2 teeth during each cycle. If the lap weight/yard is 820 grains/yd and if the comber extracts 10% noil, find the time taken to consume these laps. Also, find the production per shift of 8 hours.

　　　[Ans: 3 hours 35 min. and 49 sec; 160.19 kg/shift]

Example No. 5: A comber was worked for 1 hour when the total sliver collected on the table was 16 kg. The noil collected at the back for eight heads was – 195 g, 205 g, 200 g. 210 g, 208 g, 196 g, 202 g and 199 g respectively. Calculate the noil% of individual heads, average noil% and range.

　　　[Ans: 8.88%, 9.29%, 9.09%, 9.5%, 9.42%, 8.92%, 9.17% and 9.04%

　　　Average noil = 9.16%; range = 0.62%]

Example No. 6: Expected hanks on the comber machine are 16.5 per shift. The actual draft is 48 whereas the draft between the feed roller and lap roller is 1.08. If the comber extracts 11% noil and runs at 90 nips/min, find the feed per nip of the feed roller.

　　　[Ans: 0.24"]

Example No. 7: A comber with six heads has laps weighing 700 grains/yd. The noil extracted is 10% whereas, the mechanical draft is 54.5. The mill wants to change the noil to 13% and yet keep the hank of sliver delivered the same. If the present working change pinion has 40^T, what should be the required teeth of this C.P. Also find the mechanical draft and the hank of the sliver delivered.

　　　[Ans: 41.37^T or 41^T; draft = 52.69; sliver hank = 0.12]

Example No. 8: A comber with eight heads has a 13^T feed roller ratchet and a feed roller of 1" diameter. The machine runs at 170 nips/min and extracts 12% noil. The mechanical draft is 60 and the lap weight fed at the back is 900 grains/yd. Find the production per shift of 8 hours at 88% efficiency and the number of hanks registered at the end of the shift. By changing the ratchet to 15^T, how much will the change be in the production? Will this affect the hanks recorded on the hank meter at the end of the shift?

[Ans: 197.82 kg/shift; 34.42 hanks/shift; production reduced by 13.3%; new expected hanks will be 29.83 per shift]

Example No. 9: The lap roller of 2.375" diameter has a ratchet of 75^T and is driven by a pawl that pushes 5^T per combing cycle. A small gear of 27^T on the ratchet short shaft drives another 63^T wheel on the lap roller shaft. Similarly, the feed roller has a ratchet of 14^T and has a diameter of 1". In one combing cycle, 1^T of feed ratchet is pushed. How much is the draft between the two? By what percentage would the calculation of production differ when calculated from feed roller instead of that from lap roller?

[Ans: Draft = 1.052; production shown higher by 5.21% than actual]

Example No. 10: There are 20 conventional slow-speed combers working in a mill. They are to be replaced by new high-speed combers. The following are the particulars:

Find the number of high-speed machines.

[Ans: 3.44 New Machines]

TABLE 9.6
Comparative data of Slow-Speed and High-Speed Combers

M/c Particulars	Slow Speed	High Speed
1) Feed/nip	0.18"	0.215"
2) Nips/min	99	185
3) No. of Heads	6	8
4) Noil	13%	13%
5) Efficiency	78%	86%
6) Lap Weight – Grains/yd	480	850

Example No. 11: A comber with six heads, runs at 100 nips/min and works with laps of 600 grains/yd. The length fed per nip is 0.19", whereas the production from the hank meter is 13 hanks per shift of 8 hours. If the sliver hank is 0.15 and efficiency is 82%, find the waste extracted and mechanical draft in the comber.

[Ans: 18.876% waste; mechanical draft = 52.59]

LITERATURE REFERRED

1. Elements of Cotton Spinning – Dr. A.R.Khare, Sai Publication, Mumbai
2. Spinning Calculations – Dr. H.V. Shreeniwas Murthy and Dr. A.R.Khare
3. Whitin Super Lap Manual
4. Whitin Comber Manual

BIBLIOGRAPHY

1. *Manual of Cotton Spinning: "Draw Frames, Comber & Speed Frames": Frank Charnley*, The Textile Institute Manchester, Butterworths, 1964
2. *Technology of Short Staple Spinning: W. Klein*, Textile Institute Manual of Textile Technology
3. *Process Control in Spinning, A.R.Garde & T.A. Subramanian, ATIRA Silver Jubilee Monographs*, ATIRA Publications, 1974
4. *Spun Yarn Technology: Eric Oxtoby, Sen. Lecturer in Yarn Manufacture*, Leicester Polytechnic, U.K. Butterworth Publication, 1987
5. *Fundamentals of Spun Yarn Technology: Carl A. Lawrence*, CRC Press, London, New York, Washington, DC.
6. *Elements of Cotton Spinning: Dr. A.R.Khare*, Sai Publication, Mumbai
7. Appropriate Technology for Textile Production & Environmental Issues: International Textile Engineering Symposium, Nov. 1984, Bombay, Jointly sponsored by India-ITEME Society Bombay, India & The Textile Institute, Manchester, U.K.
8. *Textile Yarns Technology, Structure & Applications – B.C.Goswami, J.G.Martindale & F.L. Scardino*, John Willey & Sons Publication, 1977
9. *The Economics, Science & Technology of Yarn Production – P. R.Lord, Abel C. Lineberger Professor of Textile, 1974-78-79-81*, School of Textiles, North Carolina State University, Raleigh, NC 27650
10. Recent Advances in Spinning Technology: International Conference, 1995, BTRA, Mumbai
11. *Spinning: Drawing, Combing & Roving: Book of Papers: Dr. R. Chattopadhyay, Dr. R.S. Rangasamy: NCUTE Programme Series*, 1999
12. *Man-made Fibres: Dr. B.L. Deopura, Dr. B. Gupta: Book of Papers*, NCUTE Programme Series, Feb. 1999
13. Book of Papers – 37th Jt. Technological Conference, 1996, ATIRA-SITRA-NITRA & BTRA
14. Comber E 7/4, Ribbon Lap E 4/1 – Rieter Spinning booklets
15. Comber LK 69 – LMW pamphlet
16. Trumac booklets
17. Uni-Comb – Nitto Shoji Ltd. Booklet
18. Super Lap Former SL 100, Comber VC 5 – Toyota booklet

Index

Whitin Comber drafts, 189
Whitin Comber general gearing, 186
Whitin Comber speeds, 186
Whitin Comber surface speed, 188
Whitin Draw Frame, 14, 26
Whitin super lap former, 26
Wider feed distance, 60
Wll-textured lap, 117

Wooden or metallic spool, 18
Wooden spool, 38

Y

Yarn Quality Factor (YQF), 122
Yarn strength, 13
Yarn tenacity, 32

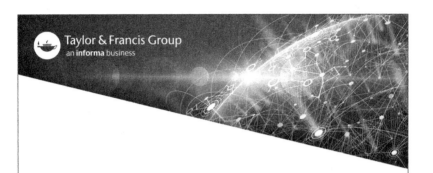

Printed in the United States
by Baker & Taylor Publisher Services